新编建筑工程专业英语

New English in Architectural Engineering

（第4版）

主　编　佟　芳
副主编　张　端　范可心

哈尔滨工业大学出版社

内 容 简 介

本书以培养学生的专业英语阅读能力为主要目标，由13个单元组成，每个单元由2篇专业阅读和会话组成，内容涉及工程造价管理、建筑材料、建筑工程经济、建筑施工、建筑工程招投标与合同管理、建筑学、高层建筑、钢筋混凝土结构、预应力混凝土结构、钢结构和工程质量管理等。为了便于读者检索和学习，本书还在书后附录部分整理出建筑工程领域常用英语词汇表和常用标识语，增强了实用性。

本书可作为高等学校工程造价管理和建筑工程专业的专业英语教材使用，也可作为具备一定英语基础的工程技术人员及自学者的参考书。

图书在版编目(CIP)数据

新编建筑工程专业英语/佟芳主编. —4版. —哈尔滨：哈尔滨工业大学出版社，2020.7(2022.1重印)
ISBN 978－7－5603－8967－7

Ⅰ.①新… Ⅱ.①佟… Ⅲ.①建筑工程-英语-高等学校-教材 Ⅳ.①TU

中国版本图书馆 CIP 数据核字(2020)第135218号

策划编辑	王桂芝
责任编辑	张 荣 陈雪巍
出版发行	哈尔滨工业大学出版社
社 址	哈尔滨市南岗区复华四道街10号 邮编150006
传 真	0451－86414749
网 址	http://hitpress.hit.edu.cn
印 刷	黑龙江艺德印刷有限责任公司
开 本	880mm×1230mm 1/32 印张7.5 字数227千字
版 次	2012年7月第1版 2020年7月第4版 2022年1月第2次印刷
书 号	ISBN 978－7－5603－8967－7
定 价	26.00元

(如因印装质量问题影响阅读，我社负责调换)

第4版前言

国家教育部颁布的《大学英语教学大纲》把专业英语列为必修课纳入英语教学计划,目的是通过专业英语课程的学习,培养学生阅读和翻译英文专业文献的能力,为其日后学习、工作中获取专业信息、掌握学科发展动态、参加国际工程技术合作奠定良好的语言基础。编者结合20多年来在建筑工程专业英语教学实践中的经验和体会,融汇建筑工程领域的发展现状,专门为建筑工程相关专业学生编写本书。本书可作为高等学校工程造价管理和建筑工程专业的专业英语教材使用,也可作为具备一定英语基础的工程技术人员及自学者的参考书。

本书以培养学生的专业英语阅读能力为主要目标,由13个单元组成,每个单元由2篇专业阅读和会话组成,内容涉及工程造价管理、建筑材料、建筑工程经济、建筑施工、建筑工程招投标与合同管理、建筑学、高层建筑、钢筋混凝土结构、预应力混凝土结构、钢结构和工程质量管理等。为了便于读者检索和学习,本书还在书后附录部分整理出建筑工程领域常用英语词汇表和常用标识语,增强了实用性。本次再版对上一版次存在的疏漏进行了修订,更新了附录1中的词汇,并增加了附录2——建筑工地常用标识语,在内容上更加完善。

此次再版由天津工程职业技术学院佟芳任主编,天津工程职业技术学院张端、范可心任副主编,王雁、陈静参编。具体编写分工为:佟芳负责编写第1、12单元;陈静负责编写第2、11单元及附录;张端负责编写第3~5、13单元;范可心负责编写第6、7单元;

王雁负责编写第 8~10 单元。全书由佟芳统稿,由国家注册造价工程师崔玉梅初审、国家注册造价工程师张国强主审。

  本书得以成稿并正式出版,要感谢哈尔滨工业大学孙爱荣老师的大力支持;还要感谢为我们仔细审稿的崔玉梅、张国强两位老师和其他有关同事的大力支持。本书在编写过程中参考了大量文献资料,在此一并表示感谢。由于编者水平有限,书中不足之处在所难免,恳请广大读者批评指正。

编　者

2020 年 6 月

CONTENTS

Unit 1 1
    Passage A Properties of Concrete 1
    Passage B Construction Cost Control 4
    Dialogue 7

Unit 2 9
    Passage A Reinforced Concrete 9
    Passage B Introduction to Reinforced Concrete 14
    Dialogue 1 21
    Dialogue 2 22
    Dialogue 3 22

Unit 3 24
    Passage A Prestressed Concrete 24
    Passage B Introduction to Prestressed Steel and
               Concrete 29
    Dialogue 33

Unit 4 35
    Passage A Responsibility for Quality 35
    Passage B Shrinkage and Creep of Concrete 40
    Dialogue 46

Unit 5 48
    Passage A Housing and Its History 48
    Passage B Loads 52

	Dialogue	61
Unit 6		64
	Passage A Structural Forms of Tall Buildings	64
	Passage B Design Criteria for Tall Buildings	74
	Dialogue	79
Unit 7		81
	Passage A Design of RC One-way Slabs	81
	Passage B Flexural Analysis of Reinforced Concrete Beams	90
	Dialogue	97
Unit 8		99
	Passage A High Strength Concrete in Prestressed Concrete Members	99
	Passage B High Tensile Steel in Prestressed Concrete Members	104
	Dialogue	112
Unit 9		114
	Passage A Strength of Reinforced Concrete	114
	Passage B Probability of Failure	123
	Dialogue	129
Unit 10		131
	Passage A Steel Structures	131
	Passage B Probabilities of Occurrence of Tornado Winds	139
	Dialogue	143
Unit 11		144
	Passage A Selecting a Framing Scheme	144
	Passage B High-rise Buildings	155

	Dialogue	162
Unit 12		164
	Passage A Determining Concrete Outlines	164
	Passage B Project Cost Control	170
	Dialogue	178
Unit 13		180
	Passage A Optimization	180
	Passage B Risk Analysis of the International Construction Project	190
	Dialogue	197
Appendix 1	建筑工程常用英语词汇	198
Appendix 2	建筑工地常用标识语	229
Reference		230

Unit 1

Passage A Properties of Concrete

Concrete is produced by mixing together cement, water, and mineral aggregates. This mixture is placed into a suitable mould, compacted, and allowed to harden. It is somewhat similar to building stone but has the advantage of being easily moulded into any suitable shape, and also of being conveniently reinforced with steel rods to improve its structural properties.

A concrete mix may be regarded as being made up of two parts, the aggregates (sand and stones) and the cement paste (that is, water and cement). The cement paste covers the surface of the stones and sand particles, building them together when it hardens. The aggregate is not altered in any way but is merely embedded firmly in a rock-like, hardened cement paste. This cementing material is formed by combining water and cement together chemically by a process called "hydration".

The chemical reaction takes place quite slowly and continues for many years. The hardened cement paste becomes harder as the hydration continues; consequently, concrete becomes harder and stronger as it grows older, provided that the temperature and moisture conditions are suitable. If, however, the concrete is allowed to dry out completely, its strength will increase no further. The longer it remains in a wet condition, the stronger it will become. Since it is also liable to attack by the weather or by chemicals, it is necessary to ensure that it becomes strong enough to withstand such attacks.

When concrete is properly made and has been allowed to harden for a sufficient length of time (usually a few days), it can resist compression forces very well and withstand quite severe weather conditions. It can resist abrasion, although not absolutely waterproof, and can offer a relatively high resistance to moisture penetration.

One of its advantages is its adaptability for a wide range of uses. It is possible to make concrete of any required strength within reasonable limits, but the stronger is made, the more is likely to cast. It would be uneconomic to use concrete of a higher strength than necessary for a particular job, but at the same time it must be strong enough to its job properly. The concrete designer is thus able to use the type of concrete which will do its job efficiently for the lowest cost.

If we are to understand how to use concrete properly, it is important that we realize what its limitations are. From the structural designer's point of view, one of the main disadvantages of concrete is its low tensile strength. That is to say, it is not able to resist forces tending to pull it apart. This may be overcome by reinforcing with steel bars at the part of a concrete structure where tensile stresses are likely to occur.

Concrete shrinks when it dries out and will expand and contract every time it is wetted and dried. This may set up tensile forces (or stresses) in the concrete and shrinkage cracking may occur. The designer can help to avoid this by specifying contraction joints in suitable place. The supervisor on the site can also help to reduce the effect of shrinkage if he ensures that the concrete is kept wet as long as possible.

Another property of concrete which may be responsible for cracking is its expansion and contraction due to heating and cooling. The designer can overcome this by including expansion and contraction joints in suitable place to allow the concrete to move freely when the temperature changes. Steel also expands when it is heated and it is very fortunate that it expands the same amount as concrete does for a given temperature rise. For this reason, steel embedded in concrete will move with the concrete without setting up any tensile stress.

The efficient use of steel bars in concrete depends on the ability of concrete to grip the steel tightly enough to prevent it from pulling out. This interaction between steel and concrete is called "bond", and the designer must check his calculations to ensure that the bond strength is sufficient. Bond may be reduced if the steel bars have loose scale, loose rust, or oil on the surface when they are embedded in the concrete. The supervisor should therefore check the steel to make sure that a good bond strength can develop.

It has already been mentioned that concrete is not entirely waterproof, but good concrete is waterproof enough for most practical purposes. It is, however, necessary to take most particular care with construction joints where there is a danger that such joints will allow water to pass through them.

Good-quality and poor-quality concrete might appear very similar by visual inspection. In order to assess the quality of the concrete properly it is necessary to use a more reliable method of inspection in the form of a suitable test. Experiments have shown that the test which generally gives the most reliable measurement of concrete quality is the crushing (or compressive) strength test. If a concrete has a high crushing strength, it will usually have good durability (that is, resistance to weather, chemical attack, and abrasion), as well as being strong enough to carry heavy structural loads. This test, which is convenient to carry out, is usually made on a cube which is crushed in a mechanical testing machine.

Words and Expressions

concrete ['kɔnkriːt]　*n.* 混凝土
cement [si'ment]　*n.* 水泥
aggregate ['ægrigit]　*n.* 骨料
embed [im'bed]　*v.* 埋置
hydration [hai'dreiʃən]　*n.* 水化作用
moisture ['mɔistʃə]　*n.* 湿气
severe [si'viə]　*a.* 不利的
abrasion [ə'breiʒən]　*n.* 磨损
penetration [ˌpeni'treiʃən]　*n.* 穿透
adaptability [ədæptə'biliti]　*n.* 适用性
tensile strength　抗拉强度
reinforce [ˌriːin'fɔːs]　*v.* 配筋
shrink [ʃriŋk]　*v.* 收缩
contraction joint　伸缩缝
site [sait]　*n.* 工地现场
grip [grip]　*v.* 抓住
bond [bɔnd]　*n.* 黏结力
scale [skeil]　*n.* 鳞状物
durability [ˌdjuərə'biləti]　*n.* 耐久性

Passage B　Construction Cost Control

Few businesses can survive without knowledge of costs and without an intelligent control of costs. Certainly, this is true in the construction industry. A contractor may be an excellent builder, but unless he knows his construction costs, he will never survive the vigorous competition in the industry. If a manufacturer finds that he has lost money on certain items, he may be able to raise the prices enough to assure a profit. However, a contractor who discovers after a project is finished that he has lost money may not have an opportunity to raise the price on the next project, especially if his losses were so great that he cannot finance another project. He may lose money because of one or more reasons, such as

1. Low bid.
2. Insufficient knowledge of job conditions.
3. Increase in the costs of materials and labor.
4. Adverse weather conditions.
5. Improper selection of construction equipment.
6. Inefficient management and supervision.

While it may not be possible to correct the first tour difficulties after the project is started, there may be some opportunity to improve item 5, and certainly an alert businessman should correct item 6, or better still he should not let it occur. Cost engineering or cost control will assist in correcting losses resulting from inefficient management and supervision. Cost control is more than mere bookkeeping. Bookkeeping will enable a contractor to determine whether he made a profit after a project is finished. Cost control during the period of construction will enable a contractor to analyze intelligently the performance of labor and equipment. It will show costs and production for labor and equipment. If the costs are higher than were estimated, either the estimate was too low or the costs are too high. If the latter condition is found to exist, it may be corrected while the project is still in operation, thereby providing a profit instead of a loss.

The owner of equipment should use an equipment ledger to

provide information concerning each type of major equipment, showing an assigned number, with a description giving the size or capacity and any auxiliary equipment, dare of purchase, name of seller, original total cost, estimated total life, and a depreciation schedule. He should use an equipment-operating ledger to keep a complete record of the cost of each type of equipment.

Prior to starting construction on a project, a contractor should set up a classification of construction accounts in which specific item numbers are assigned to each construction operation. The item numbers that were used in estimating the cost of the project should be used in preparing the classification of construction accounts. This procedure will facilitate the comparison of casts with the original estimates. In setting up the items for which costs are to be estimated and reported during construction, it is well to consider the desirability of dividing an operation into subitems. For example, the costs of concrete in a structure might be subdivided into the costs of producing aggregate, hauling aggregate, mixing and placing concrete, and finishing and curing concrete. If a concrete structure includes various sizes and shapes whose costs vary considerably, it may be desirable to divide the project into subitems for cost purposes.

Cost accounts should provide for the showing of the costs of materials, labor, and equipment separately for each operation if they are to serve the purpose for which they are used. Some contractors follow the practice of grouping the cost of all equipment into one item. This practice is not good, as it does not permit a determination of the true complete cost of a given operation on which the equipment is used. This is especially true of engineering construction for which the cost of equipment may represent a major portion of the total cost. If the cost of equipment. includes rental or depreciation, maintenance and repairs, fuel, supplies, etc. , a record of the time that the equipment is used on each operation will permit the total cost to be prorated correctly between the several operations. It is not correct to charge to an operation the cost of major repairs because the equipment was assigned to that operation when the repairs were made.

Cost accounting methods should be realistic, simple, and understandable. They are not an end product, but a means of managing a project. If the men who are supposed to use the information understand it, they will use it. If the information is too complicated, it will be disregarded or used incorrectly.

Words and Expressions

bid [bid]　　*n.* 出价,投标
supervision [ˌsjuːpəˈviʒən]　　*n.* 监督,管理
subitem [sʌbaitəm]　　*n.* 分项

Dialogue

A: Let's see the Organization Chart first.
B: Total round 200. The top management has three people only. They are Project Manager, Deputy Project Manager and Chief Engineer. Under project management there are five departments Works, Technical, Procurement, Administration and Quality Departments.
A: What about the functions of each department?
B: The Works Department is the largest one and responsible for all construction sections and teams. The Deputy Project Manager is also the Manager of Works Department. The Technical Department is in charge of all matters about drawings and designing of temporary facilities.
A: Which department deals with the subcontract matters?
B: It also belongs to the responsibility of the Technical Department.

The Procurement Department is in charge of material and construction equipment supply and warehouse. The Administration Department is in charge of all maters related to finance and public relationship. The Quality Department is a special agency for it is controlled by not only Project Management but also the Headquarter of the company.

Words and Expressions

Organization Chart　组织机构表
Project Manager　项目经理
Deputy Project Manager　项目副经理
Chief Engineer　总工程师

Unit 2

Passage A Reinforced Concrete

Mild steel has a high tensile strength, this being about 200 times that of concrete. It can be embedded in concrete to carry any tensile stress which may occur, such as in beams and slabs which are subjected to bending. Let us consider a plain concrete beam, that is, one having no steel embedded in it. If this beam is supported at its ends and loaded along its lengths, it may break, as shown in Fig. 2.1. Such a failure would occur suddenly and the beam would collapse completely.

You will notice in Fig. 2.1(a) that the underside of this beam is being pulled by tension stresses while the top side is being pushed by compressive stresses. The concrete is strong enough to carry the compression at top, but is not strong enough to carry the tension at the bottom.

If a number of steel rods were embedded in the concrete near the bottom of the beam, the beam would be able to carry a much greater load before breaking, as the steel rods would have to be broken before the beam could collapse. The steel rods would not prevent the

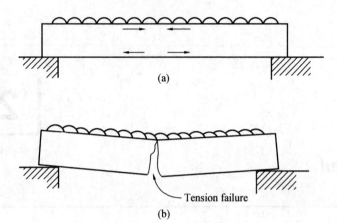

Fig. 2.1 Failure of a concrete beam with no reinforcement

concrete from cracking, and, in fact, most reinforced concrete beam are cracked when they are carrying the load for which they were designed. This is illustrated in Fig. 2.2.

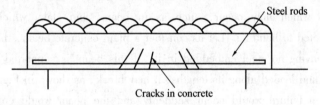

Fig. 2.2 Concrete beam reinforced with steel rods

The cracks so formed need not worry us since the steel rods would be quite capable of holding the two halves together safely.

There may, however, be a danger that moisture would enter the cracks and cause the steel to rust. The designer of the beam must ensure that the width of any crack formed is not large enough to allow corrosion of the steel to occur in this way.

In some beams diagonal cracking may occur due to a tendency of the middle part of the beam to drop out, as shown in Fig. 2.3. Such cracks are sometimes called shear cracks, and to avoid this, the designer may specify the inclusion of vertical stirrups or bent-up bars similar to those illustrated in Fig. 2.4.

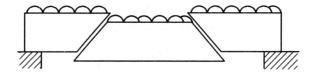

Fig. 2.3 Failure of a beam by shear cracking

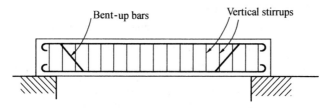

Fig. 2.4 Beam reinforced to resist shear cracking

The best position of the steel in beam will depend on how the beam is supported as well as on the loading it is expected to carry. For example, a cantilevering beam bends so that its top surface is in tension and its bottom surface is in compression. The main reinforcement must therefore be embedded near the top surface of the beam (Fig. 2.5) and cracks may occur near the support at the top.

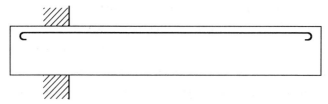

Fig. 2.5 A reinforced cantilevering beam

Generally, the design of reinforced concrete is the responsibility of the engineer, who sends instructions and drawings to the supervisor on the site. If these instructions are carried out exactly as required by the engineer, it is not really essential for the site personnel to understand the reasons for the instructions. On the other hand, if the site supervisor has a clear understanding of the elementary principles involved in reinforced concrete design, he can interpret the drawings and instructions more intelligently and should find it easier to cooperate with the engineer. This is particularly so in appreciating the need for accuracy in positioning steel reinforcement.

  Let us consider the steel in a reinforced concrete beam. If the steel is placed too near the outside of the beam, the thickness of the concrete covering the steel may become too small. It has been found by experience that a certain minimum amount of cover is necessary so that the steel is protected from moisture and prevented from rusting. If the concrete cover is too small the steel will rust and this rust will push the concrete cover off, exposing the steel to even more severe corrosion. We must not, however, place the steel too far from the surface, since this may have effect of weakening the beam in another way. It can see from figure that steel has its greatest effect in strengthening the beam if it is placed as near to the bottom surface as possible. If we were to place the steel half-way between the top and bottom surface of the beam it would not help to strengthen the beam very much. The engineer's drawings will show the best position for the steel so that the concrete cover is sufficient to protect it. His calculations for the strength of the beam will be based on the assumption that the steel will be placed accurately, and he will expect the co-operation of the site personnel in ensuring that this is done.

  There are three main types of steel used in concrete. The commonest type consists of ordinary mild-steel bars which vary in diameter from 3/16 in. ① to 2 in. , although bars greater than 1 in. diameter are only used in large scales construction work.

---

  ①   1 in. ≈2.54 cm。

High-tensile steel bars are also used for ordinary reinforced concrete and are of similar diameter to mild-steel bars. This type of steel is normally allowed to carry one and half times the load which mild steel of the same diameter can carry.

High-tensile wires and alloy-steel bars used entirely for prestressed concrete, which will be dealt within the next passage, have much higher strengths than the steels used for ordinary reinforced concrete. The allowable stresses used for these high-tensile wires may be up to about eight times that for mild steel and the stress for the high-tensile alloy bars is about four and half times the mild-steel stress.

Steel for ordinary reinforced concrete may be obtained in a variety of forms apart from the normal round bars. These include deformed and indented bars and a variety of various type of twisted steel. These bars can usually be relied upon to provide a better bond with the concrete than that obtained with ordinary round bars. In addition to the improved bond, some of these types have a higher tensile strength and can be allowed to carry a greater load.

Mild steel has a high ductility compared with most of the high-strength steels, and this enables it to stretch without breaking when it is suddenly overloaded beyond its ordinary safe working load. This also enables it to be bent readily into the required shapes.

The type of high-tensile steel used for ordinary reinforced concrete is less ductile than mild steel and is more difficult to bend, but it is generally quite suitable for bending into the shapes normally required.

Words and Expressions

slab [slæb]　　*n.* 板
collapse [kə'læps]　　*v.* 塌落
corrosion [kə'rəuʒən]　　*n.* 腐蚀

shear crack   剪切裂缝
inclusion [in'kluːʒən]   n. 内含物
stirrup ['stirəp]   n. 箍筋
cantilevering beam   悬臂梁
reinforcement [ˌriːin'fɔːsmənt]   n. 钢筋
appreciate [ə'priːʃieit]   v. 理解,评价
weaken ['wiːkən]   v. 削弱,降低
indented bar   刻痕钢筋
twisted steel   螺纹钢筋
ductility [dʌk'tiliti]   n. 延性
stretch [stretʃ]   v. 伸长
readily ['redili]   adv. 容易地

Passage B   Introduction to Reinforced Concrete

General

Reinforced concrete is one of the structural materials that is commonly used all over the world. Its two component materials, concrete and steel, work together to form structural members that can resist many types of loadings. The key to its performance lies in strengths that are complementary: concrete resists compression and steel reinforcement resists tension forces.

Although reinforced concrete is a heterogeneous complicated material, theoretical and experimental work makes the analysis and ensign of structural members, to some extent, straightforward and easy. It finds wide application in buildings and other structures because of its durability, high resistance to static and dynamic loads, resistance to fire and weathering, the availability of raw materials, and low maintenance costs. Structures such as bridges, water tanks, factories, tunnels, dams, viaducts, roads, pavings, retaining walls, and foundations can be constructed from reinforced concrete. It also has a long service life because the strength of concrete actually increases with time, provided that the steel bars are protected from corrosion.

The joint behavior of steel and concrete is based on the following properties:

1. A bond is maintained between steel and concrete after the concrete hardens. The use of deformed steel bars greatly improves the bonding, and both materials deform together under load.

2. The coefficients of thermal expansion of concrete (10×10^{-6} to $15 \times 10^{-6}/°C$) and steel ($12 \times 10^{-6}/°C$) are very close. Under changes of temperature not exceeding 80 °C, differential strains are not observed and slipping of the steel bars is not expected.

3. Concrete protects the steel reinforcement against corrosion and improves the fire resistance of the whole structural member.

Concrete is made of cement, aggregate and water, proportioned in such a way as to produce the required designed strengths.

Historical Background

Concrete was used by the ancient Romans in the construction of walls and roofs, but concrete has its first modern record as early as 1760, when in Britain John Smeaton used it in the first lock on the river Calder. The walls of the lock were made of stones filled in with concrete. In 1796, J. Parkr rediscovered the Roman natural cement, and 15 years later Vicat burned a mixture of clay and lime to produce cement. In 1824, Joseph Aspdin manufactured Portland cement in Wakefield, Britain. It was called Portland cement because when hardened it resembled stone from the quarries of the Isle of Portland.

In Britain, reinforced concrete was used in 1832 by Sir Marc Isambard Brunel in an experimental arch. In France, Francois Marten

Le Brun built a concrete house in 1832 in Moissac, in which he used concrete arches of 18-foot spans. He used concrete to build a school in St. Aignan in 1834 and a church in Corbariece in 1835. Joesph Louis Lambot exhibited a small rowboat made of reinforced concrete at the Paris Exposition of 1854. In the same year, W. B. Wilkinson of England obtained a patent for a concrete floor rein-forced by twisted cables. The Frenchman Francois Coignet obtained his first patent in 1855 for the system he used of iron bars, embedded in concrete floors, that extended to the supports. One year later, he added nuts at the screw ends of the bars, and in 1869 he published a book describing the applications of reinforced concrete.

Many investigators give credit for the invention of reinforced concrete to the gardener Joseph Monier, who obtained his patent in Paris on July 16, 1867. He made garden tubs and pots of concrete reinforced with iron mesh, which he exhibited in Paris in 1867. In 1873, he registered a patent to use reinforced concrete in tanks and bridges, and four year later, he registered another parent to use it in beams and columns.

Thaddeus Hyatt, a lawyer, was a pioneer in experimenting with reinforced concrete in the United States. He conducted flexural tests on 50 beams that contained iron bars as tension reinforcement and published the results in 1877. In other experiments, he exposed reinforced concrete slabs to high temperature to explore their fire resistance. Investigating the shrinkage of concrete and coefficients of expansion of concrete and steel, he found that both concrete and steel can be assumed to behave in a homogeneous manner for all practical purposes. This assumption was important for the design of reinforced concrete member using elastic theory. He used prefabricated slabs in his experiments and considered that prefabricated units were best cast of T-sections placed side by side to from a floor slab. Hyatt is generally credited with developing the principles upon which the analysis and design of reinforced concrete is based.

The first reinforced concrete house in the United States was built by W. E. Ward near Port Chester, New York. It used reinforced concrete for walls, beams, slabs, and staircases. The use of cavity walls was suggested by Ward to protect the house from storms, fire, and humidity. P. B. Write wrote in *The American Architect and Building News* in 1877, describing the applications of reinforced concrete in Ward's house as a new method in building construction. Ward gave a lecture on concrete and steel as structural materials, suggesting the use of reinforced concrete on a large scale.

In San Francisco, E. L. Ransome, head of the Concrete-Steel Company, used reinforced concrete in 1870 and deformed bars for the fire time in 1884. During 1889～1891, he built the two-story Leland Stanford Museum in San Francisco using reinforced concrete. He also built a reinforced concrete bridge in San Francisco in addition to some industrial buildings for the Pacific Coast Borax Company. In 1900, after Ransome introduced the reinforced concrete skeleton, the thick wall system started to disappear in construction. He registered the skeleton type of structure in 1902 and used it in the Kelly and Jones four-story factory in Greensburg, Pennsylvania. He used spiral reinforcement in the columns as was suggested by Armand Considere of France.

A. N. Talbot, at the University of Illinois, and F. E. Turneaure and M. O. Withey, of the University of Wisconsin, conducted extensive tests on concrete to determine its behavior, compressive strength, and modules of elasticity.

In Germany, G. A. Wayss bought the French Monier patent in 1897 and published his book on Monier methods of construction in 1887. Rudolph Schuster bought the patent rights in Austria, and the name of Monier spread in Europe. This is the main reason for crediting Monier as the inventor of reinforced concrete.

France played a major role in the development of reinforced concrete thanks to Francois Hennebique, who established an engineering office and employed many engineers to design thousands of reinforced concrete structures and develop reinforced concrete as a practical structural material. Hennebique conducted many experiments on reinforced concrete beams and used stirrups and bent bars for the first time to resist the shearing forces in a structural member. He issued a monthly magazine on reinforced concrete in Paris in 1898. He used 3-meter cantilever beams in a theatre in Morges, Switzerland, in 1899 and designed reinforced concrete helical staircases for the Grand and Petit Palais in Paris in 1898, using spans of 11 meters and cantilevers of 3.6 meters.

In 1900, the Ministry of Public Works in France called for a committee headed by Armand Considere, Chief Engineer of roads and bridges, to establish specifications for reinforced concrete, which were published in 1906.

Reinforced concrete was further refined by introducing some precompression in the tension zone to decrease the excessive cracks. This was the preliminary introduction of partial and full prestressing. In 1928, Eugene Freyssinet established the practical technique of using prestressed concrete.

From 1915 to 1935, research was conducted on axially loaded columns and creep effects on concrete; in 1940, eccentrically loaded columns were investigated. Ultimate-strength design started to receive special attention, in addition to diagonal tension and prestressed concrete. The American Concrete Institute Code specified the use of ultimate-strength design in 1963 and included this method in the 1971, 1977, and 1983 codes. Building codes and specifications for the design of reinforced concrete structures are established in most counties, and research continues on developing new applications and more economical designs.

Advantages and Disadvantages of Reinforced Concrete

Reinforced concrete, as a structural material, is widely used in many types of structures.

It is competitive with steel if economically designed and executed.

The advantages of reinforced concrete can be summarized as follows:

1. It has a relatively high compressive strength.

2. It has better resistance to fire than steel.

3. It has a long service life with low maintenance cost.

4. In some types of structures, such as dams, piers, and footings, it is the most economical structural material.

5. It can be cast to take the shape required, making it widely used in precast structural components.

6. It yields rigid members with minimum apparent deflection.

The disadvantages of reinforced concrete can be summarized as follows:

1. It has a low tensile strength of about one-tenth of its compressive strength.

2. It needs mixing, casting, and curing, all of which affect the final strength of concrete.

3. The cost of the forms used to cast concrete is high. The cost of from material and workmanship may equal the cost of concrete placed in the forms.

4. It has a low compressive strength as compared to steel (the ratio is about 1 : 10, depending on the materials), which leads to

large sections in columns of multistory buildings.

5. Cracks develop in concrete due to shrinkage and the application of live loads.

Design Philosophy and Concepts

The design of a structure may be regarded as the process of selecting the proper materials and proportioning the different elements of the structure according to state-of-the-art engineering science and technology. In order to fulfill its purpose, the structure must meet the conditions of safety, serviceability, economy, and functionality. Two design concepts of reinforced concrete members are discussed in this book: the strength design and alternate design methods.

The strength design method is based on the ultimate strength of structural members assuming a failure condition, whether due to crushing of the concrete or to the yield of the reinforcing steel bars. Although there is some additional strength in the bars after yielding (due to strain hardening), this additional strength is not considered in the analysis of reinforced concrete members. In the strength design method, the actual loads, or working loads, are multiplied by load factors to obtain the ultimate design loads. The load factors represent a high percentage of the factor of safety required in the design. The ACI Code emphasizes this method of design.

The alternate design method is also called the working stress design or the elastic design method. The design concept is based on elastic theory, assuming a straight-line stress variation along the depth of the concrete section. The actual loads or working loads acting on the structure are estimated and members are proportioned on the basis of certain allowable stresses in concrete and steel. The allowable stresses are fractions of the crushing strength of concrete f'_c and the yield strength of steel f_y. For example, the allowable concrete strength in flexure is equal to 0.45 of the ultimate strength of concrete f'_c. The allowable stress for steel with a yield strength of 40 ksi (thousand pounds per square inch)① is equal to 20 ksi. Therefore, an adequate factor of safety is maintained.

Words and Expressions

reinforced concrete　钢筋混凝土
complementary [ˌkɔmpli'mentəri]　*a.* 补充的,互补的
heterognenous [ˌhetərə'dʒiːniəs]　*a.* 异质的,不均的,多样的,不纯的
straightforward [ˌtreit'fɔːwəd]　*a.* 简单的,明确的
weathering ['weðəriŋ]　*n.* 风蚀
viaduct ['vaiədʌkt]　*n.* 高架桥
retaining [ri'teiniŋ]　*n.* 挡土墙
aggregate ['æɡriɡit]　*n.* 骨料
flexural ['flekʃurəl]　*a.* 弯曲的
homogeneous [ˌhɔmə'dʒiːniəs]　*a.* 同类的,同族的
prefabricate [ˌpriː'fæbrikeit]　*v.* 预制
shearing force　剪力
cantilever ['kæntiˌliːvə]　*n.* 悬臂
helical ['helikəl]　*a.* 螺旋的
eccentrically [ik'sentrikəli]　*ad.* 偏轴地
pier [piə]　*n.* 支柱

Dialogue 1

A: Hello, Tianjin No.1 Construction Engineering Company.
B: May I speak to Mr. Liu Ming, please?

---

① ksi,机械强度单位,1 ksi = 1 千磅力/平方英寸 ≈ 6.895 N/mm² = 6.895 MPa。

A: Hold the line, please. He'll be with you in a minute.
C: Hello., this is Liu Ming speaking.
A: Hello., Mr. Liu, this is John Smith. I am telephoning to tell you that I'm arriving in Tianjin on March, 16th.
C: Which flight are you taking?
A: I'll be on board CAAC Flight 268. It is due in Tianjin at 10:20 a.m.
C: All right, I'll meet you at the airport.
A: Thank you, goodbye.
C: Goodbye.

Dialogue 2

A: Excuse me, are you Mr. John Smith from London?
B: Yes, I'm John Smith from London Hartley Petrochemical Plant.
A: I'm Liu Ming from Tianjin No. 1 Construction Engineering Company.
B: Glad to meet you, Mr. Liu. Thank you for coming to meet me.
A: It's my pleasure. Did you have a nice flight?
B: Yes, I enjoyed it very much.
A: Let me help you with your luggage. We'll take a taxi to the hotel and you can get a good rest after your long flight.
B: Thank you.

Dialogue 3

A: Good morning, Mr. Smith.
B: Good morning, Mr. Liu.
A: Did you have a good sleep last night?
B: Yes, I feel quite refreshed this morning.
A: I'm glad to hear that. We plan to hold a meeting to discuss the project today. What time is convenient for you?

B: How about nine o'clock this morning? Does that suit you?
A: Yes, that's all right. Let's meet at my office. Until then, good-bye.
B: Good-bye.

Words and Expressions

hold the line   请稍候(电话用语)
be on board   搭乘
be due in/at   到达
Hartley Petrochemical Plant   哈特利化工厂
refreshed [ri'freʃt] v. 使恢复,使振作

Unit 3

Passage A Prestressed Concrete

Mild-steel bars are used in ordinary reinforced concrete beams to overcome the tensile weakness of concrete. Another way of overcoming the weakness is by prestressing the concrete so as to induce in its compression stresses before it is subject to the normal tensile stresses due to bending. When the beam is loaded and bending occurs, the bending tensile stress is neutralized by the prestressing compression so that no actual tensile stress develops. In this way the concrete is used more efficiently and no tensile cracks occur as they do in ordinary reinforced concrete. This principle is illustrated in Fig. 3.1.

The prestressing may be done in several different ways, and these are discussed later. Usually the prestress is achieved by using high-tensile steel wires or alloy-steel bars. This steel are about two and a half to five times as strong as ordinary mild steel. Generally less concrete and much less steel are required for a prestressed concrete beam compared with a reinforced concrete beam of the same strength. This means that, for prestressed concrete, the weight of the structure will be less than if it were built of ordinary reinforced concrete. It is

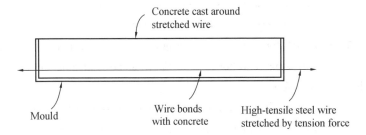

(a) Wire tensioned before concrete is placed around it

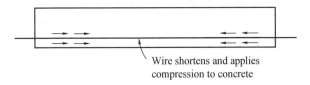

(b) When concrete is strong enough, outside wires cut off

Fig. 3.1　Pre-tensioning

necessary, however, for the quality of the concrete to be much high than that used for ordinary concrete construction and special techniques are required for the prestressing operations.

Prestressed concrete structural members are usually more springy than ordinary reinforced concrete, and behave rather more like timber in having a high degree of resilience. In some circumstances the technique is only economic when used in the form of factory-built precast units, but in many cases the prestressing can be applied successfully to in-situ concrete, particularly where large spans are concerned, and where the transporting of large units is difficult or impossible.

The concrete required for prestressed work must have a higher strength than that used for ordinary reinforced concrete, but this has the advantage that the concrete is more durable. The concrete is subjected to high compressive stresses and these are generally greatest when the prestressing is first applied to the concrete. This means that the concrete undergoes a kind of test loading before it is used to perform the job for which it was designed, and if it can withstand the initial prestressing conditions, it is likely to be well able to carry its design load safely.

Prestressing has been applied to almost all types of structure, but the most usual application is for beams. It is also particularly suitable in the construction of precast pipes and for large circular water tanks, since it is able to provide a concrete wall quite free from cracks.

There are two distinctly different methods of prestressing, one suitable for factory products and the another for use in the field. Both methods use high-tensile steel to apply prestress to the concrete, but in one case the steel is tensioned before the concrete hardens (pre-tensioned), and in the other it is tensioned after the concrete hardens (post-tensioned).

Pre-tensioning is suitable for factory-produced units and is usually done by a "long-line" process. A series of moulds is laid in a long line and a single group of wires is stretched form one end of the line to the other, so that it passes through the ends of all the moulds. Concrete is then placed into the moulds around the stretched wires and, after compacting, is allowed to harden. When the concrete has reached the necessary strength the moulds are removed and the wires protruding from the ends of the concrete units are cut off lever with the concrete. When they have been cut, the wires become shorter as they attempt to return to their original unstretched length. They are not able to shorten to their original length, since they are firmly crimped by the concrete, but they do shorten slightly and in doing so exert a compressive stresses on the concrete. This is illustrated in Fig. 3.2.

Post-tensioning techniques are of more direct interest in those engaged on site work, since it may be necessary to apply some of

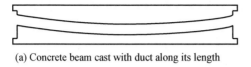

(a) Concrete beam cast with duct along its length

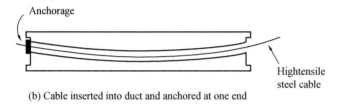

(b) Cable inserted into duct and anchored at one end

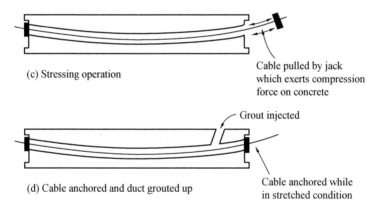

(c) Stressing operation

(d) Cable anchored and duct grouted up

Fig. 3.2   Post-tensioning

these techniques in practice.

Several systems in use are for applying tension to the steel and for providing anchorages to maintain the tension permanently. Whichever system is used, the steel is required to slide freely inside the concrete when the prestress is being applied. For this reason post-tensioning is often referred to a "non bonded" method.

In order to prevent the steel from bonding with the concrete during the prestressing process, the steel wires or rods are either inserted into preformed ducts in the concrete after this has hardened or are laid in the formwork with suitable sheathing around them so that when the concrete is placed it will not come into contact with the steel.

The cable is anchored at one end and a jack is applied to the other end. This jack grips the cable and pulls it with the required prestressing force, while at the same time pushing against the end of the concrete. While the cable is in the stretched condition, its end is anchored permanently to the concrete and the jack is removed. The final operation is the injection of cement grout into the cable duct to protect the cable from any possible corrosion. The operations involved in post-tensioning are illustrated in Fig. 3.2.

The prestressing operation must be done carefully and according to the engineer's instructions. These instructions will include the amount of force to be exerted on the cable and also the amount by which it must be stretched.

When the cable is very long or when it is placed in a duct with sharp bends the stressing is done from both ends by two jacks operated simultaneously. The amount of stretch required for a long cable may be greater than the extending capacity of a single jack, so that two are needed.

Whenever the stressing is being done the area behind the jack should be regarded as a danger zone to be kept clear in case the cable breaks suddenly. A considerable amount of energy is stored up in the cables when they are perstressed, and if they break, this energy is released suddenly and violently.

Words and Expressions

prestressed concrete   预应力混凝土

neutralize ['njuːtrəlaiz] v. 抵消
springy ['spriŋiː] a. 有弹力的
resilience [ri'ziliəns] n. 回弹
precast [ˌpriː'kɑːst] v. & a. 预浇铸(的),预制(的)
durable ['djuərəbl] a. 耐久性的
undergo [ˌʌndə'gəu] v. 经受
distinctly [dis'tiŋktli] adv. 明显地
pre-tensioning [pri'tenʃəniŋ] n. 先张法
post-tensioning [pəust'tenʃəniŋ] n. 后张法
exert [ig'zəːt] v. 施加(力)
anchorage ['æŋkəridʒ] n. 锚具,锚固
nonbonded method 无黏结法
sheathing ['ʃiːðiŋ] n. 覆盖物,保护层
cable ['keibl] n. 钢索
anchor ['æŋkə] n. 锚
jack [dʒæk] n. 千斤顶
grout [graut] n. (水泥)浆

Passage B Introduction to Prestressed Steel and Concrete

Basic Concepts of Prestressing

Prestressed concrete is basically concrete in which internal stresses of a suitable magnitude and distribution are introduction so that the stresses resulting from external loads are counteracted to a desired degree. In reinforced concrete members, the prestress is commonly introduced by tensioning the steel reinforcement.

The earliest examples of wooden barrel construction by force-fitting of metal bands and shrink-fitting of metal tyres on wooden wheels indicate that the art of prestressing has been practiced from ancient times. The tensile strength of plain concrete is only a fraction of its compressive strength and the problem of it being deficient in tensile strength appears to have been the driving factor in the development of the composite material known as "reinforced concrete".

The development of early cracks in reinforced concrete due to non-compatibility in the strains of steel and concrete was perhaps the starting point for the development of a new material like "prestressed concrete". The application of permanent compressive stress to a material like concrete, which is strong in compression but weak in tension, increases the apparent tensile strength of that material, because the subsequent application of tensile stress must first nullify the compressive prestress. In 1904 Freyssinet attempted to substitute permanently acting forces in concrete to resist the elastic forces developed under loads and this idea was later developed under the name of "prestressing".

Need for High Strength Steel and Concrete

The significant observations which resulted from pioneering research work on prestressed concrete were

1. Necessity of using high strength steel and concrete.
2. Recognition of losses of prestress due to various causes.

The early attempts to use mild steel in prestressed concrete were not successful since the working stress in mild steel of 120 N/mm^2 is more or less completely lost due to elastic deformation, creep and shrinkage of concrete.

The normal loss of stress in steel is generally about 100 to 240 N/mm^2 and it is apparent that if this loss of stress is to be a small portion of the initial stress, the stress in steel in the initial stages must be very high, about 1 200 to 2 000 N/mm^2. These high stress ranges are possible only with the use of high strength steel.

High strength concrete is necessary in prestressed concrete since the material offers high resistance in tension, shear, bond and bearing. In the zone of anchorages, the bearing stresses being higher, high strength concrete is invariably preferred to minimize costs. High strength concrete is less liable to shrinkage cracks, and has a higher modulus of elasticity and smaller ultimate creep strain, resulting in a smaller loss of prestress in steel. The use of high strength concrete results in a reduction in the cross sectional dimensions of prestressed concrete structural elements. With reduced dead weight of the material, longer spans become technically and economically practicable.

Advantages of Prestressed Concrete

Prestressed concrete offers great technical advantages in comparison with other forms of construction, such as reinforced concrete and steel. In the case of fully prestressed members, free from tensile stresses under working loads, the cross-section is more efficiently utilized when compared with a reinforced concrete section which is cracked under working loads. Within certain limits, a permanent dead load may be counteracted by increasing the eccentricity of the prestressing force in a prestressed structural element, thus effecting savings in the use of materials.

Prestressed concrete members posses improved resistance to shearing forces, due to the effect of compressive prestress, which reduces principal tensile stress. The use of curved cables, particularly in long span members helps to reduce the shear forces developed at the support sections.

A prestressed concrete flexural member is stiffer under working loads than a reinforced concrete member of the same depth. However, after the onset of cracking, the flexural behaviour of a prestressed member is similar to that of a reinforced concrete member.

The use of high strength concrete and steel in prestressed members results in lighter and slender members than could be possible by using reinforced concrete. The two structural features of prestressed concrete, namely high strength concrete and freedom from cracks, contributes to the improved durability of the structure under aggressive environmental conditions. Prestressing of concrete improves the ability of the material for energy absorption under impact loads. The ability to resist repeated working loads has been proved to be as good in prestressed as in reinforced concrete.

The economy of prestressed concrete is well established for long span structures. According to Dean, standardized precast bridge beams between 10 and 30 m long and precast prestressed piles have proved to be economical than steel and reinforced concrete in the United States. According to Abeles, precast prestressed concrete is economical for floors, roofs and bridges of spans up to 30 m and for cast in situ work, it applies to spans up to 100 m. In the long span range, prestressed concrete is generally economical in comparison with reinforced concrete and steel construction.

Prestressed concrete has considerable resilience due to its capacity for completely recovering from substantial effects of overloading without undergoing any serious damage. Leonhardt points out that in prestressed concrete elements, cracks which temporarily develop under occasional overloading will close up completely when the loads are removed. Since the fatigue strength of prestressed concrete is comparatively better than that of other materials, chiefly due to the small stress variations in prestressing steel, it is recommended for dynamically loaded structures, such as railway bridges and machine foundations. Due to the utilization of concrete in the tension zone, a saving of 15 to 30 percent in concrete is possible in comparison with reinforced concrete. The savings in steel are even higher, 60 to 80 percent, mainly due to the high permissible stresses allowed in the high tensile wires. Although there is considerable saving in the quantity of materials used in prestressed concrete members in comparison with reinforced concrete members, the economy in cost is not that significant due to the additional costs incurred for the high strength concrete high tensile steel, anchorages, and other hardware required for the production of prestressed members. However, there is an overall economy in using prestressed concrete, as the decrease in dead weight reduces the design loads and the cost of foundations.

Words and Expressions

prestressed concrete   预应力混凝土
nullify ['nʌlifai]   v. 抵消, 取消
eccentricity [ˌeksen'trisiti]   n. 偏心(距, 度)

Dialogue

A: Today we are going on discuss the tender price. From our calculation your staff cost is rather high. The monthly salary is about USD 1 500. Generally speaking, your price is very high compared with the other bidders.

B: The price of this project is worked out on the basis of our experience for power plant construction. We have reduced our profit to a very low level. Considering the benefit of the Employer, we believe that the timed completion of the works with good quality at a reasonable price is more significant than the delay of power producing. Nobody would expect a problematic situation just for a little lower price. In fact the construction cost of the civil works compared with the benefit from the earlier power producing is much smaller.

A: Why can't you do a good job with a competitive price?

B: There is an old Chinese saying that you can not let a horse run fast without feeding enough grass.

A: How many man-hours do you estimate to spend for this project?

B: About 8.9 million man-hours in minimum. So the labor cost is only USD 5 per day which is rather low. Nearly all labor force will be employed locally. The item 2 is the staff cost. It is around USD 1.9 million on total.

Words and Expressions

staff cost 人工费
benefit from 受益于

Unit 4

Passage A Responsibility for Quality

It is generally considered that responsibility rests with the contractor to construct the buildings in strict accordance with adequate plans and specifications as prepared by the architect and structural engineer. Frequently, compliance with the plans and specifications is stipulated in the specifications or the contract, or in both, to be the legal responsibility for the con-tractor. Reputable contractors, just as professional architects and engineers, conscientiously endeavor to deliver to the owner a building that will give economical, satisfactory performance as designed and specified. In seismic areas it is particularly important that all requirements of plans and specifications be observed meticulously and that all workmanship be of high quality, because the possibility of earthquake damage is greatly increased by any construction deficiencies. It is a truism of building construction that good workmanship cannot be written into the specifications. Good workmanship is the result of employment of skilled, conscientious workmen working under vigilant and constant supervision by competent foremen and by the contractor's superintendent.

Supervision

It is not sufficient to leave supervision of workmanship and other job operations entirely to the contractor and his organization. The structural engineer, who has had the responsibility for determination of site conditions and the preparation of plans and specifications for the structural parts of a building to resist the effects of earthquake motions, should be retained to supervise the construction in the interest of the owner, the safety of the public, the professional reputation of the architect and engineer, and also the reputation of the contractor. In a paper concerning the performance of structures in the Kern County earthquake it was said, "To construct earthquake resistant structures, the prime requisite is to provide for adequate engineering services both in the design and in the field supervision. Neither portion of the service is adequate alone. Too many engineers feel that if the design is adequate, the field construction will take care of itself. Experience has proved that nothing could be further from the truth."

Although the engineer as agent for the owner does not guarantee the work of the contractor, and in no way relieves the contractor of his responsibilities under the contract of which the plans and specifications are a part, the engineer must endeavor by general supervision to guard the owner against defects and deficiencies in the work. When in the engineer's judgment the intent of the plans and specifications is not being followed and he has been unable to secure compliance by the contractor, the owner should be notified so that appropriate measures may be taken to ensure compliance.

Supervision by the structural engineer should include continuous on-site inspection during the construction of all structural parts of the building by one or more competent, technically qualified, and experienced inspectors employed by the owner on the recommendation of the structural engineer. The inspectors should be under the structural engineer's supervision and direction and should report directly to him or through the resident engineer, if the project is large enough to require one. It is the responsibility of the inspector to be sure that all details of the structural engineer's design drawings, and shop drawings and bar-placing plans when provided and approved by the structural engineer, are constructed exactly as shown, that all requirements of the specifications are met, and that workmanship and construction practices are of high standard. The structural engineer should make sure that all mechanical and electrical installations required are thoroughly coordinated with the structural design so that the strength and stiffness of members will not be affected unless taken into account in the design. The inspector should make sure as the job progresses that the mechanical and electrical installations are in accordance with drawings approved by the structural engineer and that any other nonstructural items do not adversely interfere with structural elements.

  Supervision by the structural engineer and by the job inspector are quite different particularly in one very important respect. The inspector should have no authority to change plans or specifications or to made his own interpretations, even though he may be and preferably should be a structural engineer with design as well as construction experience. If any question of interpretation arises, if there is a disagreement of understanding between the inspector and the contractor, or if any possibility of error or deviation from good practice should be noticed, it should immediately be brought to the attention of the structural engineer for decision. Professional supervision, on the other hand, includes authority to modify the plans and specifications consistent with the contracts, between the owner and the engineer (sometimes through an architect) and between the owner and the contractor, if job conditions indicate a change would be in the interest of improvement of the structure or otherwise justified and consistent with the sound design principles followed in the original design.

Inspector's Authority

Certain specific authority should be given to the inspector in order that he may function effectively; the contractor should be made cognizant of that authority by the structural engineer. The inspector should be authorized to:

1. Prohibit concreting until all preliminary preparations have been made and approved, including construction of forms and placing and securing of reinforcement, inserts, pipes, and other items that are to be embedded in the concrete.

2. Forbid the use of materials, equipment, or methods that have not been approved by the structural engineer or do not conform to specifications or will result in improper construction or inferior workmanship.

3. Stop any work that is not being done in accordance with the plans and specifications.

4. When specifically authorized by the structural engineer, require the removal or repair of faulty construction or of construction completed without inspection that cannot be inspected subsequently.

Stopping work or requiring the removal of completed work and reconstruction should be very carefully considered before being ordered, but should not be avoided if the safety of the structure is involved or the quality of the work is definitely inferior to that specified.

It is essential that the job inspector and those in responsible charge of work for the contractor are experienced in concrete construction and have a thorough knowledge of the fundamentals of high-quality concrete. Attention, however, will be directed in this manual to construction details and practices considered particularly significant in the construction and performance of buildings in seismic areas, even though some may be common to good construction in general. It should not be construed that factors and practices not specifically mentioned are unimportant. For further information a list of references includes manuals and standards of the American Concrete Institute, standards of the American Society for Testing Materials, and selected publications of the Concrete Reinforcing Steel Institute, the National Ready Mixed Concrete Association, the Portland Cement Association, and the Wire Reinforcement Institute. The inspector should be familiar with and should have available for use of the job at such references.

Words and Expressions

restrict [ri'strikt]   v. 限制,约束
contractor ['kɔnˌtræktə]   n. 承包人
strict [strikt]   a. 严格的
compliance [kəm'plaiəns]   n. 符合
reputable ['repjətəbl]   a. 有信誉的
conscientiously [kɔnʃi'enʃəsli]   adv. 谨慎地
seismic ['saizmik]   a. 地震的
meticulously [mə'tikjuləsli]   adv. 小心地
deficiency [di'fiʃənsi]   n. 缺陷
truism ['truːizəm]   n. 明明白白的事
employment [im'plɔimənt]   n. 雇用
vigilant ['vidʒilənt]   a. 留心的
foreman ['fɔːmən]   n. 工长
competent ['kɔmpitənt]   a. 称职的

workmanship [ˈwəːkmənʃip] n. 工艺
requisite [ˈrekwizit] a. 必要的
adversely [ˈædvəːsli] adv. 相反地
deviation [ˌdiːviˈeiʃən] n. 失误
cognizant [ˈkɔgnizənt] a. 认知的,合理的
removal [riˈmuːvəl] a. 可移动的
construe [kənˈstruː] v. 解释

Passage B Shrinkage and Creep of Concrete

Shrinkage

The change in the volume of drying concrete is not equal to the volume of water removed. The evaporation of free water causes little or no shrinkage. As concrete continues to dry, water evaporates and the volume of the restrained cement paste changes, causing concrete to shrink, probably due to the capillary tension that develops in the water remaining in concrete. Emptying of the capillaries causes a loss of water without shrinkage. But once the absorbed water is removed, shrinkage occurs.

Many factors influence the shrinkage of concrete caused by the variations in moisture conditions.

1. Cement and water content. The more cement or water content in the concrete mix, the greater the shrinkage.

2. Composition and fineness of cement. High-early-strength and low-heat cements show more shrinkage than normal portland cement. the finer the cement, the greater the expansion under moist conditions.

3. Type, amount, and gradation of aggregate. The smaller the size of aggregate particles, the greater the shrinkage. The greater the aggregate content, the smaller the shrinkage.

4. Ambient conditions, moisture and temperature. Concrete specimens subjected to moist conditions undergo an expansion of 200×10^{-6} to 300×10^{-6}, but if they are left to dry in air, they shrink. High temperature speeds the evaporation of water and consequently increases shrinkage.

5. Admixtures. Admixtures that increase the water requirement of concrete increase the shrinkage value.

6. Size and shape of specimen. Under drying conditions, shrinkage of large masses of concrete varies from the surface to the interior, causing different shrinkage values. This will cause internal tensile stresses, and cracks may develop.

7. Amount and distribution of reinforcement. As shrinkage takes place in a reinforced concrete member, tension stresses develop in the concrete and equal compressive stresses develop in the steel. These stresses are added to those developed by the loading action. Therefore, cracks may develop in concrete when a high percentage of steel is used. Proper distribution of reinforcement, by producing better distribution of tensile stresses in concrete, can reduce differential internal stresses.

The values of final shrinkage for ordinary concrete vary between 200×10^{-6} and 700×10^{-6}. For normal weight concrete, a value of 300×10^{-6} may be used. The British Code CP110 gives a value of 500×10^{-6}, which represents an unrestrained shrinkage of 1.5 mm in 3 m length in thin plain concrete sections. If the member is restrained, a tensile stress of about 10 N/mm^2 (1 400 psi[①]) arises. If concrete is kept moist for a certain period after setting, shrinkage is reduced; therefore it is important to cure the concrete for a period of no fewer

---

① psi,磅力/平方英寸,是一种压强单位,1 psi≈0.006 895 MPa。

than 7 days.

Exposure of concrete to wind increases the shrinkage rate on the upwind side. Shrinkage cause an increase in the deflection of structural members, which, in turn, increases with time. Symmetrical reinforcement in the concrete section may prevent curvature and deflection due to shrinkage.

Generally, concrete shrinks at a high rate during the initial period of hardening, but at later stages the rate diminishes gradually. It can be said that 15 to 30 percent of the shrinkage value occurs in 2 weeks, 40 to 80 percent in 1 month, and 70 to 85 percent in 1 year.

Creep

Concrete is an elastoplastic material, and beginning with small stresses, plastic strains develop in addition to elastic ones. Under sustained load, plastic deformation continues to develop over a period that may last for years. Such deformation increases at a high rate during the first 4 months after application of the load. This slow plastic deformation under constant stress is called creep.

The instantaneous deformation is ε_1, which is equal to the stress divided by the modulus of elasticity. If the same stress is kept for a period of time, an additional strain ε_2, due to creep effect, can be recorded. If load is then released, the elastic strain ε_1 in addition to some creep strain will be recovered. The final permanent plastic strain ε_3 will be left. In this case $\varepsilon_3 = (1-\alpha)\varepsilon_2$, where α is the ratio of the recovered creep strain to the total creep strain. The ratio α ranges between 0.1 and 0.2. The magnitude of creep recovery varies with the previous creep and depends appreciably upon the period of the sustained load. Creep recovery rate will be less if the loading period is increased, probably due to the hardening of concrete while in a deformed condition.

The ultimate magnitude of creep varies between 0.2×10^{-6} and 2×10^{-6} per unit stress (lb/in.^{2}①) per unit length. A value of 1×10^{-6} can be used in practice. The ratio of creep strain to elastic strain may be as high as 4.

Creep takes place in the hardened cement matrix around the strong aggregate. It may be attributed to slippage along planes within the crystal lattice, internal stresses caused by changes in the crystal lattice, and gradual loss of water from the cement gel in the concrete.

The different factors that affect the creep of concrete can be summarized as follows:

1. The level of stress. Creep increases with an increase of stress in specimens made from concrete of the same strength and with the same duration of load.

2. Duration of loading. Creep increases with the loading period. About 80 percent of the creep occurs within the first 4 months, 90 percent after about 2 years.

3. Strength and age of concrete. Creep tends to be smaller if concrete is loaded at a late age. Also, creep of 2 000 psi (14 N/mm^2) strength concrete is about 1.41×10^{-6}, while that of 4 000 psi(28 N/mm^2) strength concrete is about 0.8×10^{-6} per until stress and length of time.

4. The ambient conditions. Creep is reduced with an increase in the humidity of the ambient air.

5. Rate of loading. Creep increases with an increase in the rate of loading when followed by prolonged loading.

6. The percentage and distribution of steel reinforcement in a reinforced concrete member. Creep tends to be smaller for higher

---

① lb/in.^{2},磅力/平方英寸。其中,lb 为力学单位"磅力",1 lb≈4.448 N; in.^{2} 为面积单位"平方英寸",1 in≈6.452 cm^2。故 1 lb/in.^{2}≈0.689 N/cm^2。

proportion or better distribution of steel.

7. Size of the concrete mass. Creep decreases with an increase in the size of the tested specimen.

8. The type, fineness, and content of cement. The amount of cement greatly affects the final creep of concrete as cement creeps about 15 times as much as concrete.

9. The water-cement ratio. Creep increases with an increase in the water-cement ratio.

10. Type and grading of aggregate. Well-graded aggregate will produce dense concrete and consequently a reduction in creep.

11. Type of curing. High-temperature steam curing of concrete as well as the proper use of a plasticizer will reduce the amount of creep. Fig. 4.1 shows the variation of creep defamation with time.

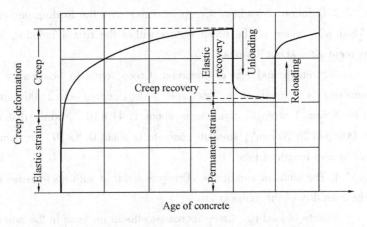

Fig. 4.1 Elastic and creep deformation of concrete under loading, release of load, and reloading conditions

Creep develops not only in compression, but also in tension, bending, and torsion. The ratio of the rate of creep in tension to that in compression will be greater than 1 in the first two weeks, but this ratio decreases over longer periods.

Creep in concrete under compression has been tested by many

investigators. Troxell, Davis, and Raphael measured creep strains periodically for up to 20 years and estimated that of the total creep after 20 years, 18~35 percent occurred in 2 weeks, 30~70 percent occurred in 3 months, and 64~83 percent occurred in 1 year.

Various expressions relating creep with time have been suggested by different investigators. One simple expression is

$$C_r = x\sqrt[y]{t} \qquad (4.1)$$

where

C_r——creep strain per unit stress per unit length;

x, y——constants to be determined by experiments;

t——time in days.

For normal concrete loaded after 28 days, $C_r = 0.13\sqrt[3]{t}$. The stress range limit due to creep as suggested by ACI Committee 215 for a minimum stress is

$$f_{cr} = 0.4f'_c - \frac{1}{2}f_{min} \qquad (4.2)$$

Creep augments the deflection of reinforced concrete beams appreciably with time. In the design of reinforced concrete members, long-term deflection may be critical and has to be considered in proper design. Extensive deformation may influence the stability of the structure.

Sustained loads affect the strength as well as the deformation of concrete. A reduction of up to 30 percent of the strength of unreinforced concrete may be expected when concrete is subjected to a concentric sustained load for 1 year.

The fatigue strength of concrete is much smaller than its static strength. Repeated loading and unloading cycles in compression lead to a gradual accumulation of plastic deformations. If concrete in compression is subjected to about 2 million cycles, its fatigue limit is about 50 to 60 percent of the static compression strength. In beams, the fatigue limit of concrete is about 55 percent of its static strength.

Words and Expressions

capillary [kə'pilən] *a.* 毛细作用的
symmetrical [si'metrikəl] *a.* 对称的
elastoplastic [i'læstə₁plæstik] *a.* 弹性塑料的
matrix ['meitriks] *n.* 基体
plasticizer ['plæstisaizə] *n.* 增塑剂,[生物][地质]基质
torsion ['tɔːʃən] *n.* 扭转
creep [kriːp] *n.* 徐变

Dialogue

A: We'd like to have a look at your temporary site facilities.
B: All right, I'll show you around. Let's first go to the production area.
A: Thank you.
B: Here we are. This is our precast yard for the reinforcement concrete pile. These two gantry cranes are set for handling of the piles.
A: What is the capacity of these cranes?
B: Each gantry crane in the production line can lift 10 t. Here is the rebar yard. The bending machine and cutting machine can process the rebar with a diameter of 40 mm.
A: I think you should have a shed for the yard. In such a hot area, if workers can work under a shed, it will certainly increase their efficiency.
B: That's a good suggestion. We'll make a shed as a roof for the rebar yard.

Words and Expressions

site inspection 现场检查
temporary site facilities 临时设施
production area 生产区

gantry crane 龙门吊
precast yard 钢筋混凝土预制场
reinforcement concrete pile 钢筋混凝土桩
bending machine 弯筋机
cutting machine 切断机
process ['prəuses] v. 加工
rebar yard 钢筋场
diameter [dai'æmitə] n. 直径
shed [ʃed] n. 遮棚

Unit 5

Passage A Housing and Its History

Housing

    Housing is living quarters for human beings. The basic function of housing is to provide shelter from the element, but people today require much more than this of their housing. A family moving into a new neighbourhood will want to know if the available housing meets its standards of safety, health, and comfort. A family may also ask how near the housing is to churches, schools, stores, the library, a movie theatre, and the community center.

    In the mid-1960's a most important value in housing was sufficient space both inside and out. A majority of families preferred single-family homes on about half an acre of land, which would provide space for spare-time activities. Many families preferred to live as far out as possible from the center of a metropolitan area, even if the wage earners had to travel some distance to their work. About four out of ten families preferred country housing to suburban housing because their chief aim was to get far away from noise, crowding, and confusion. The accessibility of public transportation had ceased to be a decisive factor in housing because most workers drove their cars to work. People were chiefly interested in the arrangement and size of rooms and the number of bedrooms.

The majority of residents in rural and suburban areas live in single-family dwellings. Housing developments containing many single-family units have been built by professional land developers. N. Y. Levittown, is a mass-produced development that contains homes for more than 60 000 people. Since they are mass-produced, development houses tend to be identical or very similar. For the same reason they are usually less expensive than houses that are individually built.

In crowded areas where land is fairly expensive, the semidetached, or two-family house is frequently found. In a semidetached house two dwellings share a central wall. Construction and heating are cheaper than in one-family house, but residents have less privacy. Row houses, in which a number of single-family houses are connected by common walls on both sides, are still less expensive to build.

City land is too expensive to be used for small housing units, except in the upper price brackets. A more efficient type of building, which houses many families on a plot of ground that would hold only a few single-family units, is the multistory multiple dwelling, or apartment house. Apartment houses may range from houses of only a few stories, without elevators, to structures of 20 or more stories, with several elevators. Some apartment houses offer city dwellers a terrace or a backyard, where they can grow a few plants or eat outdoors. Many apartment houses provide Laundromats, garage space, and gardened foyers. Huge apartment developments may cover several square blocks and include parks, playgrounds, shops, and community centers.

Two modern ideas in apartment housing are cooperative and condominium apartments. In cooperative housing all tenants together form a corporation. Each tenant owns the right to occupy his apartment by virtue of having bought shares in the corporation and by paying his share of charges for building maintenance, service, and repair. In case a tenant defaults on taxes or mortgage payments, the corporation as a whole is responsible. Condominium housing differs in that each individual actually owns and holds title to his apartment and accepts sole financial responsible for it. The individual tenant also pays a share of the maintenance of such public facilities in the buildings as elevator, hallways, and incinerators.

History of Housing

In ancient times, housing developed largely without any central planning or control. Many towns and cities were encircle by fortified walls for military protection. Urban dwellings, even including the houses of the rich, tended to be closely crowded together within the walls. In the countryside the typical community was the village, often along row of small huts or cottages in which peasant farmer lived. Landowning nobles often held country estates. In the Middle Ages some of the greater nobility lived in large fortified castles with courtyards in which the peasants could find protection in case of attack. As the countryside became more orderly, the wealthy built handsome unfortified houses surrounded by extensive parks.

With the coming of the Industrial Revolution, the cities expanded rapidly to accommodate the built by speculators who saw a chance for a quick profit. In the absence of zoning or building restrictions, they often built poorly planned cheap housing that quickly deteriorated into slums. In the United States, vast slum areas developed in the larger cities, especially in Chicago and New York. Somewhat more substantial housing was built by industrial companies for their employees. Textile towns, such as Lowell, Mass, and mining towns in Kentucky and Pennsylvania were company housing communities. As a rule, row houses made up the streets of the company towns. The houses were drearily identical, ill lighted, and often unsanitary.

In great Britain some steps to improve housing conditions were taken by humanitarian and charitable groups, such as the Society for improving the dwelling of the Labouring Classes, which was formed in 1845. Government entered the field in 1851 with set minimum standards for lower-class housing.

In the United States the dangerous and unsanitary conditions of slum living gave rise to the first tenement house regulations. which were passed in New York City in 1867 and revised and strengthened in 1879 and again in 1901. These laws set minimum standards in such matters as light, ventilation, fire protection, and santiaiton. Laws patterned on the New York City code sprang up in many other parts of the country. With the Great Depression of the 1930's came a shift in emphasis in housing laws from merely regulating the conditions of housing to providing government aid for the building of low-cost homes. The Federal Housing Administration (FHA) was established in 1934 to administer a program of government insurance of loans for building houses. Three years later the Wagner-Steagall Act set up the U. S. Housing Authority. It was empowered to lend up 90 percent of the cost of approved projects to clear slums and build low-income family housing. In 1947 the functions of a number of housing agencies were absorbed by the Housing and Home Finance Agency (HHFA), which was replaced in 1965 by the department of Housing and Urban Development.

Words and Expressions

housing [ˈhauziŋ]   n. 住房, 住房建筑
quarters [ˈkwɔːtə(r)z]   n. 住处
metropolitan [ˌmetrəˈpɔlitən]   a. 主要都市的, 大城市的
semidetached [ˌsemidiˈtætʃt]   a. (房屋)一侧与邻屋相连的, 半独立的
bracket [ˈbrækit]   n. 等级, 阶层
laundromat [ˈlɔːndrəumæt]   n. 洗衣店(间)
foyer [ˈfɔiei]   n. (剧场, 旅馆等处的)门厅, 休息处
condominium [ˌkɔndəˈminiəm]   n. (公寓中)个人拥有的一套房间, 共管
mortgage [ˈmɔːgidʒ]   n. 押借款
hallway [ˈhɔːlˌwei]   n. 厅, 过道
incinerator [inˈsinəreitə]   n. 垃圾焚烧炉
by virtue of   凭借, 依靠
to hold title to sth.   持某物的权利
fortified wall   城墙
influx [ˈinflʌks]   n. 流入, 汇集
unsanitary [ˌʌnˈsænitəri]   a. 不卫生的, 不清洁的
substantial [səbˈstænʃəl]   a. 结实的, 坚强的
ventilation [ˌventiˈleiʃən]   n. 通风

Passage B   Loads

The accurate determination of the loads to which a structure or structural element will be subjected is not always predictable. Even if the loads are well-known at one location in a structure, the distribution of load from element to element throughout the structure usually requires assumptions and approximations. Some of the most common kinds of loads are discussed in the following sections.

Dead Load

Dead load is a fixed position gravity service load, so called because it acts continuously toward the earth when the structure is in service. The weight of the structure is considered dead load, as well as attachments to the structure such as pipes, electrical conduit, air-conditioning and heating ducts, lighting fixtures floor covering, roof covering, and suspended ceilings; that is, all items that remain throughout the life of the structure.

Dead loads are not usually known accurately until the design has been completed. Under steps 3 through 6 of the design procedure discussed in Sec. 1.2, the weight of the structure or structural element must be estimated, preliminary section selected, weight recomputed, and member selection revised if necessary. The dead load of attachments is usually known with reasonable accuracy prior to the design.

Tab. 5.1 Typical Minimum Uniformly Distributed Live Loads

Occupancy or Use	Live load	
	psf[①]	Pa
1. Hotel guest rooms, School classrooms, Private apartments, Hospital private rooms	40	1 900
2. Offices	50	2 400
3. Assembly halls, fixed seat Library reading rooms	60	2 900
4. Corridors, above first floor in schools, libraries, and hospitals	80	3 800

① psf,磅力/平方英尺,1 psf≈47.88 Pa。

Continued 5.1

Occupancy or Use	Live load	
	psf	Pa
5. Assembly areas; theater lobbies Dining rooms and restaurants Office building lobbies Main floor, retail stores Assembly hall, movable seats	100	4 800
6. Wholesale stores, all floors Manufacturing, light Storage warehouses, light	125	6 000
7. Armories and drill halls Stage floors Library stack rooms	150	7 200
8. Manufacturing, heavy Sidewalks and driveways subject to trucking Storage warehouse, heavy	250	12 000

Live Load

Gravity loads acting when the structure is in service, but varying in magnitude and location, are termed live loads. Examples of live loads are human occupants, furniture, movable equipment, vehicles, and stored goods. Some live loads may be practically permanent, others may be highly transient. Because of the unknown nature of the magnitude, location, and density of live load items, realistic magnitudes and the positions of such loads are very difficult to determine.

Because of the public concern for adequate safety, live loads to be taken as service loads in design are usually prescribed by state and local building codes. These loads are generally empirical and conservative, based on experience and accepted practice rather than accurately computed values. Wherever local codes do not apply, or do not exist, the provisions from one of several regional and national building codes may be used. One such widely recognized code is the

American National Standard Mihijum Design Loads for Buildings and Other Structures ANSI A58. 1 of the American National Standards (ANSI), from which some typical live loads are presented in Tab. 5. 1. The code will henceforth be referred to as the ANSI Standard. This Standard is updated from time to time, most recently in 1982.

Live load when applied to structure should be positioned to give the maximum effect, including partial loading, alternate span loading, or full span loading as may be necessary. The simplified assumption of full uniform loading everywhere should be used only when it agrees with reality or is an appropriate approximation. The probability of having the prescribed loading applied uniformly over an entire floor, or over all floors of a building simultaneously, is almost nonexistent. Most codes recognize this by allowing for some percentage reduction from full loading. For instance, for live loads of 100 psf or more ANSI standard allows members having an influence area of 400 sq · ft[①] or more to be designed for a reduced live load according to Eq. 5. 1, as follows

$$L = L_0 \left[0.25 + \frac{15}{\sqrt{A_I}} \right] \qquad (5.1)$$

where

L ——reduced live load per sq ft of area supported by the member;

L_0——unreduced live load per sq ft of area supported by the member (from Tab. 5. 1);

A_I——influence area, sq ft.

The influence area A_I area is four times, the tributary area for a column two times, the tributary area for a beam, and is equal to the

---

① sq · ft, square feet,平方英尺,1 sq · ft≈0.092 9 m²。

panel area for a two-way slab. The reduced live load L shall not be less than 50% of the live load L_0 for members supporting one floor, nor less than 40% of the live load L_0 otherwise.

The live load reduction referred to above is not permitted in areas to be occupied as places of public assembly and for one-way slabs, when the live load L is 100 psf or less. Reductions are permitted for occupancies where L_0 is greater than 100 psf and for garages and roofs only under special circumstances (ANSI-4.7.2).

Snow Load

The live loading for which roofs are designed is either totally or primarily a snow load. Since snow has a variable specific gravity, even if one knows the depth of snow for which design is to be made, the load per unit area of roof is at best only a guess.

The best procedure for establishing snow load for design is to follow the ANSI Standard. This Code uses a map of the United States giving isolines of ground snow corresponding to a 50-year mean recurrence interval for use in designing most permanent structures. The ground snow is then multiplied by a coefficient that includes the effect of roof slope, wind exposure, nonuniform accumulation on pitched or curved roofs, multiple series roofs, and multilevel roofs and roofs areas adjacent to projections on a roof level.

It is apparent that the steeper the roof the less snow can accumulate. Also partial snow loading should be considered, in addition to full loading if it is believed such loading can occur and would cause maximum effects. Wind may also act on a structure that is carrying snow load. It is unlikely, however, that maximum snow and wind loads would act simultaneously.

In general, the basic snow load used in design varies from 30 to 40 psf (1 400 to 1 900 MPa) in the northern and eastern states to

20 psf (960 MPa) or less in the southern states. Flat roofs in normally warm climates should be designed for 20 psf (960 MPa) even when such accumulation of snow may seem doubtful. This loading may be thought of as due to people gathered on such a roof. Furthermore, though wind is frequently ignored as a vertical force on a roof, nevertheless it may cause such an effect. For these reasons a 20 psf(960 MPa) minimum loading, even though it may not always be snow, is reasonable. Local codes, actual weather conditions, ANSI, or the Canadian Structural Design Manual, should be used when designing for snow. Other snow load information has been provided by investigators.

Wind Load

All structures are subject to wind load, but they are usually only those more than three or four stores high, other than long bridges, for which special consideration of wind is required.

On any typical building of rectangular plan and elevation, wind exerts pressure on the windward side and suction on the leeward side, as well as either uplift or downward pressure on the roof. For most ordinary situations vertical roof loading from wind is neglected on the assumption that snow loading will require a greater strength than wind loading does. This assumption is not true for southern climates where the vertical loading, due to, wind must be included. Furthermore, the total lateral wind load, windward and leeward effect, is commonly assumed to be applied to the windward face of the building.

In accordance with Bernoulli's theorem for an ideal fluid striking an object, the increase in static pressure equals the decrease in dynamic pressure, or

$$q = \frac{1}{2}\rho v^2 \qquad (5.2)$$

where q is the dynamic pressure on the object, p is the mass density of air (specific weight $w=0.07651$ pcf[1] at sea level and 15 ℃), and v is the wind velocity. In terms of velocity v in miles per hour, the dynamic pressure q (psf) would be

$$q = \frac{1}{2}\left(\frac{0.07651}{32.2}\right)\left(\frac{5280v}{3600}\right)^2 = 0.0026\,v^2 \qquad (5.3)$$

In design of usual types of buildings, the dynamic pressure q is commonly converted into equivalent static pressure p, which may be expressed

$$p = qC_eC_gC_p \qquad (5.4)$$

where C_e is a exposure factor that varies from 1.0 (for 0 ~ 40-ft height) to 2.0 (for 740 ~ 1 200-ft[2] height); C_g is a gust factor, such as 2.0 for structural members and 2.5 for small elements including cladding; and C_p is a shape factor for the building as a whole. Excellent details of application of wind loading to structures are available in the ANSI Standard and in the National Building Code of Canada.

The commonly used wind pressure of 20 psf, as specified by many building codes, corresponds to a velocity of 88 miles per hour (mph) from Eq. 5.3. An exposure factor C_e of 1.0, a gust factor C_g of 2.0, and a shape factor C_q of 1.3 for an airtight building, along with a 20 psf equivalent static pressure p, will give from Eq. 5.4 a dynamic pressure q of 7.7 psf, which corresponds, using Eq. 5.3 to a wind velocity of 55 mph. For all buildings having nonplanar surfaces, plane surfaces inclined to the wind direction or surfaces having significant openings, special determination of the wind forces should be

---

[1] pcf,磅/立方英尺,1 pcf≈0.016 g/cm³。
[2] ft,英尺,1 ft≈0.305 m。

made using such sources as the ANSI Standard, or as the National Building Code of Canada. For more extensive treatment of wind loads, the reader is referred to the Task Committee on Wind Force, Lew, Simiu, and Ellingwood in the Building Structural Design Handbook, and others.

Earthquake Load

An earthquake consists of horizontal and vertical ground motions, with the vertical motion usually having much smaller magnitude. Since the horizontal motion of the ground causes the most significant effect, it is that effect which is usually thought of as earthquake load. When the ground under an object (structure) having a certain mass suddenly moves, the inertia of the mass tends to resist the movement. A shear force is developed between the ground and the mass. Most building codes having earthquake provisions require the designer to consider a lateral force CW that is usually empirically prescribed. The dynamics of earthquake action on structures is outside the scope of this text, and the reader is referred to Chopra, and Clough and Penzien.

In order to simplify the design process, most building codes contain an equivalent lateral force procedure for designing to resist earthquakes. One of the most widely used design recommendations is that of the Structural Engineers Association of California (SEAOC), the latest version of which is 1974. Since that time, the Applied Technology Council (ATC) prepared a set of design provisions. Some recent rules for the equivalent lateral force procedure are those given by the ANSI Standard. In ANSI the lateral seismic forces V, expressed as follows, are assumed to act nonconcurrently in the direction of each of the main axes of the structure

$$V = ZIKCSW \qquad (5.5)$$

where

Z——seismic zone coefficient, varying from 1/8 for the zone of lowest seismicity, to 1 for the zone of highest seismicity;

I——occupancy importance factor, varying from 1.5 for buildings designated as "essential facilities", and 1.25 for buildings where the primary occupancy is for assembly for greater than 300 persons, to 1.0 for usual buildings;

K——horizontal force factor, varying from 0.67 to 2.5, indicating capacity of the structure to absorb plastic deformation (low values indicate high ductility) the seismic coefficient, equivalent to the maximum acceleration in terms of acceleration due to gravity

$$C = \frac{1}{15\sqrt{T}} \leqslant 0.12 \qquad (5.6)$$

T——fundamental natural period, i.e., time for one cycle of vibration, of the building in the direction of motion;

S——soil profile coefficient, varying from 1.0 rock to 1.5 for soft to medium-stiff clays and sands;

W——total dead load of the building, including interior partitions.

When the natural period T cannot be determined by a rational means from technical data, it may be obtained as follows for shear walls or exterior concrete frames utilizing deep beam or wide piers, or both

$$T = \frac{0.05 h_n}{\sqrt{D}} \qquad (5.7)$$

where D is the dimension of the structure in the direction of the applied forces, in feet, and h_n is the height of the building.

Once the base shear V has been determined, the lateral force must be distributed over the height of the building.

More details of the ANSI Standard procedure are available in the

Building Structural Design Handbook. Various building code formulas for earthquake-resistant design are compared by Chopra and Crux. Many states have adopted the Uniform Building Code (UBC), the most recent version of which is 1985, which contains provisions for design to resist earthquake generally based on the ANSI Standard.

Words and Expressions

distribution [ˌdistri'bju:ʃən] n. 分配,分布
conduit ['kɔndit] n. 管道,导线管
duct [dʌkt] n. (输送)管道
transient ['trænziənt] a. 无常的,短期的
empirical [em'pirikəl] a. 经验的
retial ['ritel] a. 零售的
panel ['pænəl] (嵌)板
isoline ['aisəulain] n. (等值)线
suction ['sʌkʃən] n. 吸收
gust [gʌst] n. 阵风
cladding ['klædiŋ] n. 覆盖,包层
nonplanar ['nɔn'pleinə] a. 非平面的
seismic ['saizmik, 'sais-] a. 地震的
nonconcurrently [nɔnkən'kʌrəntli] adv. 非共点地
windward side 迎风面
leeward side 背风面
soil profile 土(壤)剖面

Dialogue

A: Another thing we would like to point out about the construction equipment is that you need at least one set of power generator in case the power supply from the city net switched off since power supply in the city is not stable.

B: OK. We will prepare two sets of power generators, one in construction area and the other in the living area.
A: Now let's have a discussion about the production area in the Chapter 6. You layout show that an area of 69 000 m^2 is required for your temporary facilities. It is too much. You should resubmit a revised one.
B: From our experience for such a large power station a piece of land like that is necessary. We planned carefully for our precast yard, concrete batching plants, material storage, rebar yard, carpentry shop, repair shop, warehouse, steel mill and so on, it seems very difficult to reduce from any of them.
A: Just a moment. I can't find where your laboratory is. It should not be omitted from your layout because it is an important facility for such a site.
B: It will certainly not. You see the block not far away the batching plant is the laboratory. It will be equipped with all necessary instruments for soil, concrete and steel tests necessary for the works.
A: The cost issues will be discussed later. Now let's turn on the Chapter 8: Estimation of Electricity and Water. We have no objection to your power consumption estimation but can not agree with you about the water consumption estimation.

Words and Expressions

power generator 发电机
power supply 供电
temporary facilities 临时设施
precast yard 预制场
concrete batching plants 混凝土搅拌站
material storage 储料仓

rebar yard 钢筋场
carpentry shop 木工房
warehouse [ˈweəhaus] 仓库
steel mill 铁件加工房
power consumption estimation 用电量估计
water consumption estimation 用水量估计

Unit 6

Passage A Structural Forms of Tall Buildings

Braced-Frame Structures

In braced frames the lateral resistance of the structure is provided by diagonal members that, together with the girders, form the "web" of the vertical truss, with the columns acting as the "chords" (Fig. 6.1). Because the horizontal shear on the building is resisted by the horizontal components of the axial tensile or compressive actions in the web members, bracing systems are highly efficient in resisting lateral loads.

Bracing is generally regarded as an exclusive steel system because the diagonals are inevitably subjected to tension from one or the other directions of lateral loading. Concrete bracing of the double diagonal form is sometimes used, however, with each diagonal designed as a compression member to carry the full external shear.

The efficiency of bracing, by being able to produce a laterally very stiff structure for a minimum of additional material, makes it an economical structural form for any height of building, up to the tallest. An additional advantage of fully triangulated bracing is that the girders usually participate only minimally in the lateral bracing action. Consequently, the floor framing design is independent of its level in the structure and, therefore, can be repetitive up the height of the building with obvious economy in design and fabrication. A major disadvantage of diagonal bracing is that it obstructs the internal planning and the location of windows and doors. Therefore, braced bents are usually incorporated internally along walls and partition lines, especially around elevator, stair, and service shafts. Another drawback is that the diagonal connections are expensive to fabricate and erect.

The traditional use of bracing has been in story-height, bay-width modules (Fig. 6.1) that are fully concealed in the finished building. More recently, however, larger external scale bracing, extending over many stories and bays, has been used to produce not only highly efficient structures, but aesthetically attractive buildings.

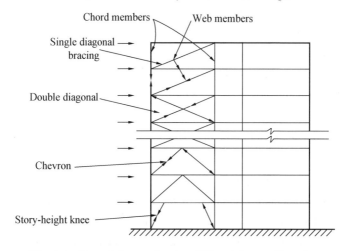

Fig. 6.1 Braced frame-showing different types of bracing

Rigid-Frame Structures

Rigid-frame structures consist of columns and girders joined by moment-resistant connections. The lateral stiffness of a rigid-frame bent depends on the bending stiffness of the columns, girders, and connections in the plane of the bent. The rigid frame's principal advantage is its open rectangular arrangement, which allows freedom of planning and easy fitting of doors and windows. If used as the only source of lateral resistance in a building, in its typical 20 ft (6 m) ~ 30 ft (9 m) bay size, rigid framing is economical only for buildings up to 25 stories. Above 25 stories, the relatively high lateral flexibility of the frame calls for uneconomically large members in order to control the drift.

Rigid-frame construction is ideally suited for reinforced concrete buildings because of the inherent rigidity of the reinforced concrete joints. The rigid-frame form is also used for steel frame building, but moment-resistant connections in steel tend to be costly. The sizes of the columns and girders at any level of a rigid frame are directly influenced by the magnitude of the external shear at that level, and they therefore increase toward the base. Consequently, the design of the floor framing can not be repetitive as in some braced frames. A further result is that sometimes it is not possible in the lowest stories to accommodate the required depth of girder within the normal ceiling space.

Gravity loading is also resisted by the rigid-frame action. Negative moments are induced in the girders adjacent to the columns causing the mid-span positive moments to be significantly less than in a simply supported span. In structures in which gravity loads dictate the design, economy in member sizes that arise from this effect tend to be offset by the higher cost of the rigid joints.

While rigid frames of a typical scale that serve alone to resist lateral loading have an economic height limit of about 25 stories, smaller scale rigid frames in the form of a perimeter tube, or typically scaled rigid frames in combination with shear walls or braced bents, can be economic up to much greater heights.

Infilled-frame Structures

In many countries infilled frames are the most usual form of construction for tall buildings of up to 30 stories. Column and girder framing of reinforced concrete, or sometimes steel, is infilled by panels of brickwork, blockwork, or cast-in-place concrete.

When an infilled frame is subjected to lateral loading, the infill behaves effectively as a strut along its compression diagonal to brace the frame. Because the infills serve also as external walls or internal partitions, the system is an economical way of stiffening and strengthening the structure.

The complex interactive behavior of the infill in the frame, and the rather random quality of masonry, has made it difficult to predict with accuracy the stiffness and strength of an infilled frame. Indeed, at the time of writing, no method of analyzing infilled frames for their design has gained general acceptance. For these reasons, and because of the fear of the unwitting removal of bracing infills at some time in the life of the building, the use of the infills for bracing tall buildings has mainly been supplementary to the rigid-frame action of concrete frames.

Flat-plate and Flat-slab Structures

The flat-plate structure is the simplest and most logical of all structural forms in that it consists of uniform slabs of 5 ~ 8 in. (12 ~ 20 cm) thickness, connected rigidly to supporting columns. The system, which is essentially of reinforced concrete, is very economical in having a flat soffit requiring the most uncomplicated formwork and, because the soffit can be used as the ceiling in creating a minimum possible floor depth.

Under lateral loading the behavior of a flat-plate structure is similar to that of a rigid frame. Its lateral resistance depends on the flexural stiffness of the components and their connections, with the slabs corresponding to the girders of the rigid frame. It is particularly appropriate for apartment and hotel construction where ceiling spaces are not required and the slab may serve directly as the ceiling. The flat-plate structure is economical for spans of up to 25 ft. (8 m), above which drop panels can be added to create a flat-slab structure for spans of up to 38 ft. (12 m).

Buildings that depend entirely for their lateral resistance on flat-plate or flat-slab action are economical up to about 25 stories. Previously, however, when Code requirements for wind design were less stringent, many flat-plate buildings were constructed in excess of 40 stories, and are still performing satisfactorily.

Shear Wall Structures

Concrete or masonry continuous vertical walls may serve both architecturally as partitions and structurally to carry gravity and lateral loading. Their very high inplane stiffness and strength makes them ideally suited for bracing tall buildings. In a shear wall structure, such walls are entirely responsible for the lateral load resistance of the building. They act as vertical cantilevers in the form of separate planar walls, and as nonplanar assemblies of connected walls around elevator, stair, and service shafts. Because they are much stiffer horizontally than rigid frames, shear wall structures can be economical up to 35 stories.

In contrast to rigid frames, the shear walls' solid form tends to restrict planning where open internal spaces are required. They are well suited, however, to hotels and residential buildings where the floor-by-floor repetitive planning allows the walls to be vertically continuous, and where they serve simultaneously as excellent acoustic and fire insulators between rooms and apartments.

If, in low-to medium-rise buildings, shear walls are combined with frames, it is reasonable to assume that the shear walls attract all the lateral loading so that the frame may be designed for only gravity loading. It is especially important in shear wall structures to try to plan the wall layout so that the lateral load tensile stresses are suppressed by the gravity load stresses. This allows them to be designed to have only the minimum reinforcement. shear wall structures have been shown to perform well in earthquakes, in which case ductility becomes an important consideration in their design.

Coupled Wall Structures

A coupled wall structure is a particular, but very common, form of shear wall structure with its own special problems of analysis and design. It consists of two or more shear walls in the same plane, or almost the same plane, connected at the floor levels by beams or stiff slabs. The effect of the shear-resistant connecting members is to cause the set of walls to behave in their plane partly as a composite cantilever, bending the common centroidal axis of the walls. This results in a horizontal stiffness much greater than if the walls acted as a set of separate uncoupled cantilevers.

Coupled walls occur often in residential construction where lateral-load resistant cross walls, which separate the apartments, consist of in-plane coupled pairs, or trios, of shear walls between which there are corridor or window openings.

Although shear walls are obviously more appropriate for concrete construction, they have occasionally been constructed of heavy steel plate, in the style of massive vertical plate or box girders, as parts of steel frame structures. These have been designed for locations of extremely heavy shear, such as at the base of elevator shafts.

Wall-frame Structures

When shear walls are combined with rigid frames, the walls, which tend to deflect in a flexural configuration, and the frames, which tend to deflect in a shear mode, are constrained to adopt a common deflected shape by the horizontal rigidity of the girders and slabs. As a consequence, the walls and frames interact horizontally, especially at the top, to produce a stiffer and stronger structure. The interacting wall-frame combination is appropriate for buildings in the 40-to 60-story range, well beyond that of rigid frames or shear walls alone.

An additional, less well known feature of the wall-frame structure is that, in a carefully "tuned" structure, the shear in the frame can be made approximately uniform over the height, allowing the floor framing to be repetitive.

Although the wall-frame structure is usually perceived as a concrete structural form, with shear walls and concrete frames, a steel counterpart using braced frames and steel rigid frames offers similar benefits of horizontal interaction. The braced frames behave with an overall flexural tendency to interact with the shear mode of the rigid frames.

Frame-tube Structures

The lateral resistance of frame-tube structures is provided by very stiff moment-resisting frames that form a "tube" around the perimeter of the building. The frames consist of closely spaced columns, 6-12 ft (2 ~ 4 m) between centers, joined by deep spandrel girders. Although the tube carries all the lateral loading, the gravity loading is shared between the tube and interior columns or walls. When lateral loading acts, the perimeter frames aligned in the direction of loading act as the "webs" of the massive tube cantilever, and those normal to the direction of the loading act as the "flanges".

The close spacing of the columns throughout the height of the structure is usually unacceptable at the entrance level. The column are therefore merged, or terminated on a transfer beam, a few stories above the base so that only a few, larger, more widely spaced columns continue to the base. The tube form was developed originally for buildings of rectangular plan, and probably its most efficient use is in that shape. It is appropriate, however, for other plan shapes and has occasionally been used in circular and triangular configurations.

The tube is suitable for both steel and reinforced concrete construction and has been used for buildings ranging from 40 to more than 100 stories. The highly repetitive pattern of the frames lends itself to prefabrication in steel, and to the use of rapidly movable gang forms in concrete, which is made rapid construction.

The framed tube has been one of the most significant modern developments in high-rise structural form. It offers a relatively efficient, easily constructed structure, appropriate for use up to the greatest of heights. Aesthetically, the tube's externally evident form is regarded with mixed enthusiasm some praise the logic of the clearly expressed structure while others criticize the grid-like facade as small-windowed and uninterestingly repetitious.

The tube structure's structural efficiency, although high, still leaves scope for improvement because the "flange" frames tend to suffer from "shear lag" This results in the mid-face "flange" columns being less stressed than the corner columns and, therefore, not contributing as fully as they could to the flange action.

Tube-in-tube or Hull-core Structures

This variation of the framed tube consists of an outer framed tube, the "hull", together with an internal elevator and service core. The hull and core act jointly in resisting both gravity and lateral loading. In a steel structure the core may consist of braced frames. whereas in a concrete structure it would consist of an assembly of shear walls.

To some extent, the outer framed tube and the inner core interact horizontally as the shear and flexural components of a wall-frame structure, with the benefit of increased lateral stiffness. However, the structure tube usually adopts a highly dominant role because of its much greater structural depth.

Bundled-tube Structures

This structural form is notable in its having been used for the Sears Tower in Chicago-the world's tallest building. The Sears Tower consists of four parallel rigid steel frames in each orthogonal direction, interconnected to form nine "bundled" tubes. As in the single-tube structure, the frames in the direction of lateral loading serve as "webs" of the vertical cantilever, with the normal frames acting as "flanges".

The introduction of the internal webs greatly reduces the shear lag in the flanges; consequently, their columns are more evenly stressed than in the single-tube structure, and their contribution to the lateral stiffness is greater. This allows columns of the frames to be spaced further apart and to be less obtrusive. In the Sears Tower, advantage was taken of the bundled form to discontinue some of the tubes, and so reduce the plan of the building at stages up the height.

Braced-tube Structures

Another way of improving the efficiency of the framed tube, thereby increasing its potential for use to even greater heights as well as allowing greater spacing between the columns, is to add diagonal bracing to the faces of the tube. This arrangement was first used in a steel structure in 1969, in Chicago's John Hancock Building, and in a reinforced concrete structure in 1985, in New York's 780 Third Avenue Building. In the steel tube the bracing traverses the faces of the rigid frames, whereas in the concrete structure the bracing is formed by a diagonal pattern of concrete window-size panels, poured integrally with the frame.

Because the diagonals of a braced tube are connected to the columns at each intersection, they virtually eliminate the effects of shear lag in both the flange and web frames. As a result, the structure behaves under lateral loading more like a braced frame, with greatly diminished bending in the members of the frames. Consequently, the spacing of the columns can be larger and the depth of the spandrels less, thereby allowing larger size windows than in the conventional tube structure.

In the braced-tube structure, the bracing contributes also to the improved the performance of the tube in carrying gravity loading differences between gravity load stresses in the columns are evened out by the braces transferring axial loading from the more highly to the less highly stressed columns.

Words and Expressions

chord [kɔːd]   *n.* (翼,桁)弦,弦杆
fabrication [ˌfæbriˈkeiʃən]   *n.* 建造,装配

aesthetically [iːs'θetikli] adv. 艺术上,美学上
rectangular [rek'tæŋgjulə] a. 矩形的
repetitive [ri'petitiv] a. 重复的
planar ['pleinə] a. 平面的
shaft [ʃɑːft] n. 竖井
insulator ['insjuleitə] n. 隔离
ductility [dʌk'tiliti] n. 延性
trio(s) ['triːəu] n. 三件一套
perimeter [pə'rimitə] n. 周长
hull [hʌl] n. 主体,外壳
orthogonal [ɔː'θɔgənl] a. 正交的,直角的
intersection [ˌintə'sekʃən] n. 交叉(点,线)
spandrel ['spændrəl] n. 上下层窗空间,窗台下的墙

Passage B   Design Criteria for Tall Buildings

The construction of tall buildings is the result of urbanization as seen in America since the late nineteenth century when the fist skyscrapers were built. The industrialized society attracts more people to the cities, requiring more space for offices as well as for habitation. Tall buildings, however, require two basic technical ingredients. First, economic method of building a tall building must be found and second, a reliable and economic method of transporting people vertically through the building must be available. Even though the Otis elevator in the late 19th century provided the logical answer to vertical transportation, the structure still remained a very significant deterrent to building very tall buildings. The pioneering Chicago School of Architecture refined the use of beam-column frames, but still required high "premium for height" for buildings taller than say 20 stories. As a result, buildings built up to 1920s were mostly below stories. A heroic step was taken in 1930 when the Empire State

Building was built with more or less conventional method of construction. This was no example of any economic breakthrough. From the time of Depression in the 30's hardly any high rise construction was undertaken. After World War II the socioeconomic needs in the U. S. opened the way for taller office and apartment buildings. But, these buildings could no longer be designed in the same manner and proportion as the earlier buildings, simply because the concept of uncluttered column-free space, the concept of climate control and communication in the buildings has drastically changed. All these changes in attitudes brought the era of a new architecture. The new demand and challenge was to create total urban environments.

Within the last few years, research on building materials such as reinforced concrete and structural steel have made great strides and opened horizons for more efficient use of these materials. The structural engineers and architects also have met the challenge to find efficient and economical new structural systems for various ranges and heights of buildings going all the way to well over 100 stories. Consequently, the process of selecting a structural system for a tall building has become more complicated than it ever was.

The process of choosing the structural system for a tall building depends on many criteria which are not always structural. Without the understanding of all of these significant criteria, the structural engineer would feel frustrated in his efforts. The process of designing a building is multi-discipline, and since structure is an integral part of the total scheme, the successful adoption of an innovation in concept will have a much harder time of for acceptance if the engineer does not have an understanding of these criteria, the following is a brief discussion of these design criteria.

1. Environmental Planning Consideration. In any environment the

addition of a tall building will certainly influence the operation of the traffic system, as well as the flow of people in the entire neighborhood. Therefore, a tall building project must be resolved in terms of pedestrian, auto and other traffic, and also resolve the overall need for space at the ground level. While an open plaza may be a viable solution under certain climatic and planning conditions, it may not be a reasonable solution where weather conditions are extreme either for winter or for summer. In such cases alternate to open plazas must be considered as a pan of the total solution, and the structural system must respond to it.

2. The Overall Proportions of the Tower. The relationship of the building to the surrounding environment and other existing buildings and plazas may dictate the proportions of the tower itself. It is, of course, obvious that a flat narrow tower will have a higher height-to-width ratio, which normally would mean more lateral sway. The increase in height-to-width ratio will generally mean increase in premium for height caused by additional material required to reduce lateral sway, as well as to increase resistance to overturning. Therefore, where slender buildings is an important requirement it is essential to find alternate structural solutions of higher efficiency even though such a system may increase fabrication and construction costs relative to more standard forms of construction.

3. Permissible Floor Area Ratio. The construction of buildings in the city area is generally controlled by zoning. Zoning, in most cases, unfortunately, is the product of political and economic consideration. Zoning would normally allow a maximum number of square feet that can be built at any given site In dense urban centers this becomes a critical consideration, particularly because in terms of high land cost the more area that can be built on a given piece of land, the more the potential for economic return. Under the free enterprise system,

therefore, there is every reason to believe that investors would like to build the fill allowable floor area ratio, which will then mean a taller building than otherwise considered consistent with the environment, Here again, the challenge to the architects and engineers is to provide viable alternatives using different structural systems from which a more rational and a more satisfying solution can be chosen.

4. Inner Space Criteria. Only a few years back, column spacing of 20 ft. was accepted as a structural limitation that could not be overcome. Newer structural systems have given the architect and developers the choice to create larger column free spaces in the office, as well as residential buildings. In the office buildings for instance, 35 ft. clear spacing is now normally considered the minimum and most developers would not mind if they can get a 60 ft. clear span between the core and the exterior walls. It is this kind of design consideration that has led to the development of a number of structural systems which do not require any column in the space between the core and the exterior walls. For apartments and hotel buildings the reverse is normally the case, that is, the maximum distance from the corridor to the outer wall does not normally exceed 30 ft. , because every room is generally required to have an exterior exposure. This kind of planning requirement means that apartment and hotel buildings cannot be wider than between 70 to 90 ft. Although wider apartment buildings have been built, the width of such buildings are still far less than what would be used for efficient office buildings, This type of requirement, of course, immediately means the possibility of a higher height-to-width ratio for residential buildings, and consequent selection of structural systems different than for office buildings.

5. Climatic Considerations. Climatic considerations sometimes play a strong part in choosing the structural system. For extremely cold winter climate the general tendency is to provide larger, clear

windows, perhaps because there is a greater need for the occupant to look out and enjoy the nature as much as possible. On the other hand, in hot tropical climate it is more economical to have less glass and more solid masonry or concrete surface. This coincides with lesser need for the occupants to look outside. Structural systems which inherently use large exterior bearing walls are, therefore, quite often welcome in such climatic conditions.

6. Structural Material Considerations. The selection of a system depends strictly on the local relative economies between the various structural materials. It is because of this reason that while, in one area a concrete structural system may be economical, and in another area of the country a structural steel system may become more economical. Sometimes, of course, a combination of both materials in the same building may result in the optimum design.

7. Foundation Considerations. Tall buildings require a stable foundation. Where rock is directly available near the ground level, the choice of structural system or material is not affected. However, in many areas where, for instance, floating foundation is the only way to support a building, the total weight of the structure becomes very significant because it controls the total depth of excavation. In such cases, light weight construction is the obvious structural choice. This could lead to either a steel structural framing, or an all lightweight concrete construction, or a composite system depending on the local material economies.

8. Time of Construction. Although time of construction of a project seems to be an obvious factor in choosing the type of construction, often it is overlooked as such, and estimates to compare structural systems are made without any consideration of the total time of construction. This may have little effect where interim financing during construction is not a critical item. In most investment projects,

however, this cannot be overlooked. If the time of construction of a 1.5 million sq. ft. building takes 6 months more than another competitive structural system, it may very well mean an equivalent loss of $1.5 million in the total construction cost. In choosing structural systems, therefore, this criterion should be considered and only eliminated, if specifically requested by the owner.

Words and Expression

coincide [ˌkəuinˈsaid] *vi.* 与……一致，相符
foundation [faunˈdeiʃən] *n.* 基础
excavation [ˌekskəˈveiʃən] *n.* 开凿，挖掘
floating foundation 浮筏基础
sq. /square 平方
ft. /foot 英尺

Dialogue

A: According to the earthwork bid package by the local architect, the bidders are required to strip top soil up to 18 inches.
B: But in my observations. 6 inches is sufficient. My second visit to the job site yesterday gave me more reasons for my opinion.
A: If so, the bidders will have to reappraise their position.
B: Yes, they have to reduce their quotations.
A: In that way, the original estimated contract price for this item may be decreased by 10%.
B: Let's obtain more details before we present it to the local architect and the owner.

Words and Expressions

earthwork 土方工程
strip the top soil 除表土

bid package 投标发包
reappraise 重新估计
quotation [kwəu'teiʃən] 报价
contract price 合同价

Unit 7

Passage A Design of RC One-way Slabs

Introduction

Reinforced concrete slabs are constructed to provide flat surfaces, usually horizontal, in building floors and roofs, and other structures. The slab may be supported by masonry or reinforced concrete walls, by columns, by reinforced concrete beams usually poured monolithically with the slab, by structural steel beams, or by the ground. The depth of the slab is usually very small compared to its span.

Types of Slabs

Slabs can be classified structurally as follows:

1. One-way slabs. If a slab is supported on two opposite sides only, it will bend or deflect in a direction perpendicular to the supported edges. The structural action is one-way, and the loads are carried by slab in the deflected short direction. This type of slab is called a one-way slab(Fig. 7.1(a)). If the slab is supported on four

sides, and the ratio of the long side to the short side is equal to or greater than 2, most of the load (about 95 percent or more) is carried in the short direction, and one-way action is considered for all practical purposes (Fig. 7.1(b)). If the slab is made of reinforced concrete with no voids, it is called a one-way solid slab.

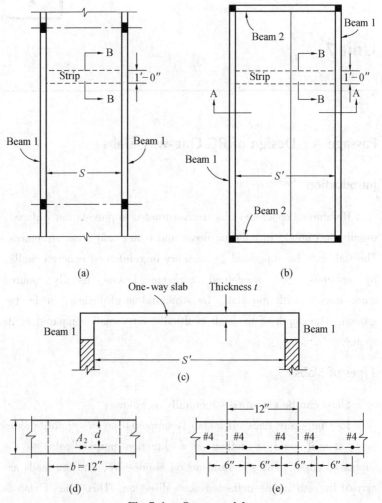

Fig. 7.1　One-way slabs

2. One-way joist floor system. This type of slab is also called a ribbed slab. It consists of a floor slab, usually 2 ~ 4 in. (50 ~ 100 mm) thick, supported by reinforced concrete ribs (or joists). The ribs are usually tapered and uniformly spaced at distances that do not exceed 30 in. (750 mm). The ribs are supported on girders that rest on columns. The spaces between the ribs may be formed using removable steel or fiberglass form fillers (pans), which may be used many times. In some ribbed slabs, the space between ribs may be filled with permanent fillers to provide a horizontal slab soffit. Different materials are used as fillers such as hollow lightweight or normal weight concrete blocks, hollow clay tile blocks, or any lightweight material.

3. Two-way Slabs. When the slab is supported on four sides and the ratio of the long side to the short side is less than 2, the slab will deflect in double curvature in both directions. The floor load is carried in two directions to the four beams and the slab is called a two-way slab.

4. Two-way ribbed slabs and the waffle slab system. This type of slab consists of a floor slab with a length-to-width ratio less than 2. The thickness of the slab is usually 2 to 4 in. (50 ~ 100 mm) and supported by ribs (or joists) in two directions. The ribs are arranged in each direction at spacings of about 20 ~ 30 in. (500 ~ 750 mm), producing square or rectangular shapes. The ribs can also be arranged at 45° or 60° from the centerline of slabs, producing architectural shapes at the soffit of the slab. In two-way ribbed slabs, different systems can be adopted:

(1) A two-way rib system with voids between the ribs is obtained by using special removable and usable form (pans) normally square in shape. The ribs are supported on four sides by girders that rest on columns. This type is called a two-way ribbed (joist) slab system.

(2) A two-way rib system with permanent fillers between ribs that produce horizontal slab soffits. The fillers may be of hollow lightweight or normal weight concrete or any other lightweight material. The ribs are supported by girders on four sides in turn supported by columns. This type is called a two-way ribbed (joist) slab system or a hollow-block two-way ribbed system.

(3) A two-way rib system with voids between the ribs has the ribs continuing in both directions without supporting beams, and resting directly on columns through solid panels above the columns. This type is called a waffle slab system.

5. Flat slabs. A flat slab is a two-way slab reinforced in two directions that usually does not have beams or girders, and the loads are transferred directly to the supporting columns. The column tends to punch through the slab, which can be treated by three methods

(1) Using a drop panel and a column capital.

(2) Using a drop panel without a column capital. The concrete panel around the column capital should be thick enough to withstand the diagonal tensile stresses arising from the punching shear.

(3) Using a column capital without drop panel.

6. Flat plate floors. A flat plate floor is a two-way slab system consisting of a uniform slab that rests directly on columns and does not have beams or column capitals. In this case the column tends to punch through the slab, producing diagonal tensile stresses. Therefore, a general increase in the slab thickness is required, or special reinforcement (as shearhead reinforcement) is used.

7. Slabs resting directly on ground. These are commonly used in basements or ground floors, sidewalks, and mostly in roads, highways, and airport runways. These slabs may be designed by empirical methods for simple cases, like basement floors, or can be analyzed by methods developed for beams on elastic foundations.

8. Lift slabs. These are specially constructed flat-plate slabs. The columns are fixed in place before casting begins. Sleeves or collars that fit loosely around the column are embedded in the concrete. The bottom level is poured first and serves as the pouring bed for the following levels until all slabs are poured. When cured, the slabs are lifted by hydraulic jacks to the desired level, where the collars are fixed to the columns.

Design of One-way Solid Slabs

As mentioned earlier, in a one-way slab the ratio of the length of the slab to its width is greater than 2. Nearly all the loading is transferred in the short direction, and the slab may be treated as beam. A unit strip of slab, usually 1 ft. (or 1 m) at right angles to the supporting girders, is considered as a rectangular beam. The beam has a unit width with a depth equal to the thickness of the slab and a span length equal to the distance between the supports. A one-way slab thus consists of a series of rectangular beams placed side by side (Fig. 7.1).

If the slab is only one span and rests freely on its supports, the maximum positive moment M for a uniformally distributed load of w psf is $M = (wL)/8$, where L is the span length between the supports. If the same slab is built monolithically with the supporting beams or is continuous over several supports, the positive and negative moments are calculated either by elastic analysis or by moment coefficients as for continuous beams. The ACI Code, section 8.3, permits the use of moment and shear coefficients in the case of two or more approximately equal spans. This condition is met when the larger of two adjacent spans does not exceed the shorter span by more than 20 percent. For uniformly distributed loads, the unit live load shall not exceed three times the unit dead load. When these conditions are not satisfied, elastic analysis is required. In elastic analysis, the negative bending moments at the centers of the supports are calculated. The value that should be considered in the design is the negative moment at the face of the support. To obtain this value, reduce from the maximum moment value at the center of the support a quantity equal to $Vb/3$, where V is the shearing force calculated from the analysis and b is the width of the support

$$M_f = M_c - \frac{Vb}{3} \qquad (7.1)$$

where

M_f——the moment at face of the support;

M_c——the moment at centerline of support.

In addition to moment, diagonal tension and development length of bars should also be checked for proper design. The ACI gave coefficients to compute the maximum and minimum moments in continuous slabs.

Design Limitations of the ACI Code

1. A typical imaginary strip 1 ft (or 1 m) wide is assumed.

2. The minimum thickness of one-way slabs using grade 60 steel according to the ACI Code, is as follows:

$L/20$ for simply supported slabs

$L/24$ for one end continuous slabs

$L/28$ for both ends continuous slabs

$L/10$ for a cantilever slab

3. Strength design as well as working stress design methods are permitted.

4. Deflection is to be checked when the slab supports are attached to construction likely to be damaged by large deflections. Deflection limits are set by the ACI Code, section 9.5(b).

5. It is preferable to choose slab depth to the nearest 1/2 in. (or 10 mm).

6. Shear should be checked, although it does not usually control.

7. Concrete cover in slabs shall not be less than 3/4 in. (20 mm) at surfaces not exposed to weather or ground.

8. In structural slabs of uniform thickness, the minimum amount of reinforcement in the direction of the span shall not be less than that required for shrinkage and temperature reinforcement (ACI Code, section 7.12).

9. The principal reinforcement shall be spaced not farther apart than three times the slab thickness, nor more than 18 in. (ACI Code, section 7.65).

10. Straight bar systems may be used in both tops and bottoms of continuous slabs. An alternative bar system of straight and bent (trussed) bars placed alternately may also be used.

11. In addition to main reinforcement, steel bars at right angles to the main rein-forcement must be provided. This additional steel is called secondary, distribution, shrinkage, or temperature reinforcement.

Reinforcement Details

In continuous one-way slabs, the steel area of the main reinforcement is calculated for all critical sections, at midspans, and at supports. The choice of bar diameter and detailing depends mainly on the steel areas, spacing requirements, and development length. Two bar systems may be adopted.

In the straight bar system, straight bars are used for top and bottom reinforcement in all spans. This is mainly recommended in slabs with small thickness, about 6 in. or less. The time to produce straight bars is less than that required to produce bent bars.

In the bent bar or trussed system, straight and bent bars are placed alternatively in the floor slab. The location of bent points should be checked for flexural, shear, and development length requirements.

Distribution of Loads from One-way Slabs to Supporting Beams

In one-way floor slab systems, the loads from slabs are transferred to the supporting beams along the long ends of the slabs. The beams transfer their loads in turn to the supporting columns.

From Fig. 7.2, it can be seen that beam B_1 carries loads from two adjacent slabs. Considering a 1-ft length of beam, the load transferred to the beam is equal to the area of a strip 1 ft wide and S ft in length multiplied by the intensity of load on the slab. This load produces a uniformly distributed load on the beam

$$U_B = U_S \times S \qquad (7.2)$$

The uniform load on the end beam B_1 is half the load on B_2, as it supports a slab from one side only.

The load on column C_4 is equal to the reaction of two beams B_2 from both ends

$$\text{Load on column} \quad C_4 = U_B L = U_S L S \qquad (7.3)$$

The load on column C_3 is one-half the load on column C_4, as it supports loads from slabs on one side only. Similarly, the load on column C_2 and C_1 are

$$\text{Load on} \quad C_2 = U_S \frac{L}{2} S = \text{Load on } C_3 \qquad (7.4)$$

$$\text{Load on} \quad C_2 = U_S \frac{L}{2} \frac{S}{2} \qquad (7.5)$$

From the above analysis, it can be seen that each column carries loads from slabs surrounding the column and up to the centerline of

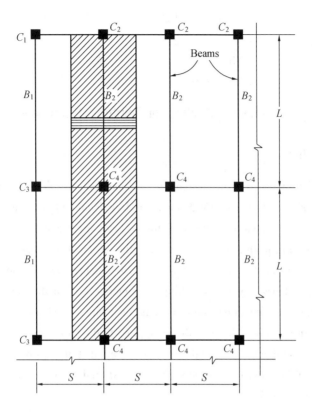

Fig. 7.2 Distribution of loads on beams

adjacent slabs $L/2$ in the long direction and $S/2$ in the short direction.

Words and Expressions

masonry ['meisənri]   n. 砌体
monolithically [ˌmɔnə'liθikli]   adv. 整体地
perpendicular [ˌpəːpən'dikjulə]   a. 垂直的
joist [dʒɔist]   n. 小梁,托梁
rib [rib]   n. 肋,梁; v. 加肋于
girder ['gəːdə]   n. 大梁

soffit ['sɔfit]   n. (拱)腹,(肋板的)下部
waffle slab   华夫板(类似于密肋板)
hydraulic jack   液压千斤顶
truss [trʌs]   n. 桁架,构架

Passage B Flexural Analysis of Reinforced Concrete Beams

Introduction

Ultimate-strength design is a method of determining the dimensions of a structural member based on ultimate loads and ultimate strengths of sections. Ultimate loads are found by multiplying the working loads, the dead load and the assumed live loads, by load factors. Ultimate strength of sections is reached by the yielding of steel, followed by the crushing of concrete. Ultimate loads cause external ultimate forces such as bending moments, shear, or thrust, depending on how these loads are applied to the structure. The section of the member is designed in such a way that its internal ultimate capacity is equal to or greater than the external ultimate forces acting on the member.

In proportioning reinforced concrete structural members, it will be necessary to investigate three main items:

1. The safety of the structure. This is maintained by providing adequate internal ultimate strength capacity.

2. Deflection of the structural member under working loads. The maximum value of this must be limited, usually specified as a factor of the span, to preserve the appearance of the structure.

3. Width of cracks under working loads. Visible cracks spoil the appearance of the structure and also permit humidity to penetrate into the concrete, causing corrosion of steel and weakening the reinforced concrete member. The ACI Code of 1983 limits crack widths for

interior and exterior exposure to 0.016 in. (0.40 mm) and 0.013 in. (0.33 mm), respectively.

It is worth mentioning that the ultimate-strength design method was first permitted in Britain in 1957, in the United States in 1956, and in the USSR in 1935. The ACI Code of 1963 put equal emphasis on both ultimate-strength design and working-stress design methods, while the ACI Codes of 1971, 1977 and 1983 emphasized ultimate-strength design.

Assumptions

Reinforced concrete sections are heterogeneous (nonhomogeneous) as they are made of two different materials, concrete and steel. Therefore proportioning structural members by ultimate-strength design is based on the following assumptions:

1. Strait in concrete is the same as in reinforcing bars at the same level, provided that the bond between the steel and concrete is adequate.

2. Strain in concrete is linearly proportional to the distance from the neutral axis.

3. The modulus of elasticity of all grades of steel is taken as
$$E_s = 29 \times 10^6 \text{ lb/in.}^2 \qquad (7.6)$$
$$(2 \times 10^5 \text{ MPa or N/mm}^2 \text{ or } 2.039 \times 10^6 \text{ kgf/cm}^2)$$

The stress in the elastic range is equal to the strain multiplied by E_s. In the plastic range, the steel stress to be considered is the yield stress f_y.

4. Plane cross sections continue to be plane after bending.

5. Tensile strength of concrete is neglected because (1) concrete's tensile strength is about 10 percent of its compressive strength, (2) cracked concrete is assumed to be not effective, and (3) before cracking, the entire concrete section is effective in resisting the

external moment.

6. The method of elastic analysis, assuming an ideal behavior at all levels of stress, is not valid. At high stresses, nonelastic behavior is assumed, which is in close agreement with the actual behavior of concrete and steel.

7. At ultimate strength, the maximum strain at the extreme compression fiber is assumed equal to 0.003, by the ACI Code provision.

8. At ultimate strength, the shape of the compressive concrete stress distribution may be assumed to be rectangular, parabolic, or trapezoidal. In this text, a rectangular shape will be assumed.

Behavior of a Simply Supported Reinforced Concrete Beam Loaded to Failure

Concrete being weakest in tension, a concrete beam under an assumed working load will definitely crack at the tension side, and the beam will collapse if tensile reinforcement is not provided. Concrete cracks occur at a loading stage when its maximum tensile stress reaches the modulus of rupture of concrete. Therefore, steel bars are used to increase the moment capacity of the beam; the steel bars resist the tensile force while the concrete resists the compressive force.

To study the behavior of a reinforced concrete beam under increasing load, let us examine how two beams were tested to failure. Both beams had a section of 4.5 by 8 in. (110 by 200 mm), reinforced only on the tension side by two No.5 bars. They were made of the same concrete mix. Beam 1 had no stirrups while beam 2 was provided with No.3 stirrups spaced at 3 inches. The loading system and testing procedure were the same for both beams. To determine the compressive strength of the concrete and its modulus of elasticity E_c, a standard concrete cylinder was tested and strain measured at different load increments. The following observations were noted at different distinguishable stages of loading.

Stage 1: At zero external load, each beam carried its own weight in addition to that of the loading system, which consisted of an I-beam and some plates. Weights of beam 1, beam 2, and the loading system were 270, 271, and 88.5 lb (1 200, 1 205, and 394 N), respectively. Both beams behaved similarly at this stage. At any section, the entire concrete section in addition to the steel reinforcement resisted bending moment and shearing forces. Maximum stresses occurred at the section of maximum bending moment, that is, at midspan. Maximum tension stress at the bottom fibers was much less than the modulus of rupture of concrete. Compressive stress at the top fibers was much less than the ultimate concrete compressive stress f_c. No cracks were seen at this stage.

Stage 2: This stage was reached when the external load P was increased from zero to P_1, which produced tensile stresses at the bottom fibers equal to the modulus of rupture of concrete. At this stage the entire concrete section was effective, with the steel bars at the tension side sustaining a strain equal to that of the surrounding concrete. Stress in the steel bars was equal to the stress in the adjacent concrete multiplied by the modular ratio n, the ratio of the modulus of elasticity of steel to that of concrete. The compressive stress of concrete at the top fibers was still as small as compared with the compressive strength f'_c. The behavior of the beams was elastic within this stage of loading.

At a load of 6 000 lb (26.7 kN), vertical cracks were seen in both beams. The stains in beams 1 and 2 were read to be 529×10^{-6} and 550×10^{-6}, which corresponds to stresses of 2 500 psi (17.24 N/mm^2) and 2 600 psi (17.9 N/mm^2), respectively.

The stain in the steel bar of beam 2 was 685×10^{-6}, which corresponds to a steel stress of 20 000 psi (138 N/mm^2) when the strain is multiplied by the modulus of elasticity of steel.

Stage 3: When the load was increased beyond P_1 (greater than 6 000 lb, 26.7 kN), tensile stresses in concrete at the tension zone increased until they were greater than the modulus of rupture f_r, and cracks developed. The neutral axis shifted upward, and cracks extended close to the level of the shifted neutral axis. Concrete in the tension zone lost its tensile strength, and the steel bars started to work effectively and to resist the entire tensile force. Between cracks, the concrete bottom fibers had tensile stresses, but of negligible value. It can be assumed that concrete below the neutral axis did not participate in resisting external moments.

In general, the development of cracks and the spacing and maximum width of cracks depend on many factors, such as the level of stress in the steel bars, distribution of steel bars in the section, concrete cover, and grade of steel used.

At this stage, the deflection of the beams increased clearly as the moment of inertia of the cracked section was less than that of the uncracked section. Cracks started about the midspan of the beam, while other parts along the length of the beam did not crack. When load was again increased, new cracks developed, extending toward the supports. The spacing of these cracks depends on the concrete cover and the level of steel stress. The width of cracks also increased. One or two of the central cracks were most affected by the load, and their cracks widths increased appreciably while the other crack widths increased much less. It is more important to investigate those wide cracks that to consider the larger number of small cracks.

If the load was released within this stage of loading. It would be observed that permanent fine cracks, of no significant magnitude, were left. On reloading cracks would open quickly, as the tensile strength of concrete had already been lost. Therefore, it can be stated that the second stage, once passed, does not happen again in the life of the beam. When cracks develop under working loads, the resistance of the entire concrete section and the gross moment of inertia are no longer valid.

At high compressive stresses, the strain of the concrete increased rapidly, and the stress of concrete at any strain level was estimated from a stress-strain graph obtained by testing a standard cylinder to failure for the same concrete. As for the steel, the stresses were still below the yield stress, and the stress at any level of stain was obtained by multiplying the strain of steel by the modulus of elasticity of steel E_s.

Stage 4: In beam 1, at a load value of 9 500 lb (42.75 kN), shear stresses at a distance of about the depth of the beam from the support increased and caused diagonal cracks at approximately 45° from horizontal in the direction of principal stresses resulting from the combined action of bending moment and shearing force. The diagonal crack extended downward to the level of the steel bars, then extended horizontally at that level toward the support. When the crack, which had been widening gradually, reached the end of the beam, a concrete piece broke off and failure occurred suddenly. The failure load was 13 600 lb (61.2 kN). Stresses in concrete and steel at the midspan section did not reach their failure stresses.

In beam 2, at a load of 11 000 lb (49.5 kN), a diagonal crack developed similar to that of beam 1, then other parallel diagonal cracks appeared and the stirrups started to take effective part in resisting the principal stresses. Cracks did not extend along the horizontal main steel bars as in beam 1. On increasing the load, diagonal cracks on the other end of the beam developed at a load of 13 250 lb (59.6 kN). Failure did not occur at this stage because of the presence of stirrups.

Stage 5: When the load on beam 2 was further increased, strains increased rapidly until the maximum carrying capacity of the beam was reached at ultimate load $P_u = 16\ 200$ lb (72.9 kN).

In beam 2, the amount of steel reinforcement used was relatively small. When the strain in the steel reached the yield strain, which can be considered equal to yield stress divided by the modulus of elasticity of steel $\varepsilon_y = f_y/E_s$, the strain in the concrete ε_c was less than the strain at maximum compressive stress f'_c. The steel bars yielded and the strain in steel increased to about as 12 times as that of the yield strain without increase in the load. Cracks widened sharply, deflection of the beam increased greatly, and the compressive strain on the concrete increased. After another very small increase of load, steel strain hardening occurred and concrete reached its maximum strain ε'_c, and it started to crush under load; then the beam collapsed.

Types of Flexural Failure

Three types of flexural of a structural member can be expected, depending on the percentage of steel used in the section.

1. Steel may reach its yield strength before the concrete reaches its maximum strength. In this case failure is due to yielding of steel. The section contains a relatively small amount of steel and is called an under-reinforced section.

2. Steel may reach its yield strength at the same time as concrete reaches its ultimate strength. The section is called a balanced section. Steel and concrete fail simultaneously.

3. Concrete may fail before the yield of steel, due to the presence of a high percentage of steel in the section. In this case the concrete strength f'_c and maximum strain ε'_c are reached while the

steel stress is less than the yield strength: $f_s < f_y$. The section is called an over-reinforced section.

It can be assumed that concrete fails in compression when the concrete strain reaches 0.003. A range of 0.002 5 to 0.004 has been obtained from tests; therefore, can be taken as 0.003.

In a structure based on under-reinforced sections, steel yields before the crushing of concrete. Cracks widen extensively, giving a warning before the concrete crushes and the structure collapses. This type of design is adopted by the ACI Code. In a structure designed with balanced or over-reinforced conditions, the concrete fails suddenly and collapse occurs immediately with no warning. This type of design is not allowed by the ACI Code. There is no doubt that to utilize steel and concrete most efficiently, a balanced section is most suitable, but due to safety needs the ACI Code limits the maximum steel percentage to 75 percent of the balanced steel ratio.

Words and Expressions

inertia [i'nəːʃjə]   n. 惯性
flexural ['flekʃərəl]   a. 弯曲的
thrust [θrʌst]   n. 推力
diagonal [dai'ægənl]   a. 对角线的,斜向的
parabolic [ˌpærə'bɔlik]   a. 抛物线状的
trapezoidal [træpi'zɔidəl] a. 不规则四边形的,梯形的

Dialogue

A: I would like to congratulate you that after our examination of your prequalification document your company has been accepted to enter the next stage for tendering of the project. Here is the Tender Document for the civil works.

B: Thank you!

A: The conditions in the Tender Document are worked out on the basis of FIDIC Conditions. As you know the FIDIC Conditions are fair to both Contractor and Employer, and we hope you will agree with these conditions.
B: We will study it carefully and be reasonable in dealing with the conditions in the document.
A: In the Tender Document, Volume 2 is the Bill of Quantities. There are 18 sections with 989 items of rates in the Volume.
B: Is the contract price a lump sum or a rates basis one?
A: This is a rates basis contract.

Words and Expressions

tender document for civil works   土建工程部分的标书
FIDIC: Conditions of Contract for Works of CMI Engineering Construction Standardized by Federation Imitational Des Ingenieurs-Conseils   菲迪克条款(国际工程师协会规定的土木工程施工合同条件)
items of rates   单价条目
lump sum contract   包干价合同
rates basis contract   单价合同

Unit 8

Passage A   High Strength Concrete in Prestressed Concrete Members

High Strength Concrete Mixes

Prestressed concrete requires concrete which has a high compressive strength at reasonably early age, with comparatively higher tensile strength than ordinary concrete. Low shrinkage, minimum creep characteristics and a high value of Young's modulus are generally deemed necessary for concrete used for prestressed members. Many desirable properties, such as durability, impermeability and abrasion resistance are highly influenced by the strength of concrete. With the development of vibration techniques in 1930, it is possible to produce high strength concrete having 28-day cube compressive strength in the range of 30 ~ 70 N/mm^2 without much difficulty. Recent developments in the field of concrete mix design have indicated that it is possible in the present state of art to produce even ultra high strength concrete, which has any desired 28-day cube compressive strength ranging from 70 ~ 100 N/mm^2, without taking recourse to unusual materials or processing and without incurring any significant technical difficulties.

Experimental investigations by Erntroy and Shacklock have indicated that in high strength concrete mixes, workability, type and maximum size of aggregate, and the strength requirement influence the selection of the water/cement ratio. Crushed rock aggregates, being angular, generally produce stronger concretes at the same age in comparison with gravel aggregates.

High strength concrete mixes can be designed by using any of the following established methods:

1. Erntroy and Shacklock's empirical method,

2. American Concrete Institute mix design procedure for no slump concrete,

3. Murdock's mix design charts based on surface and angularity index of aggregates,

4. Road Note No. 4. Procedure.

The author has demonstrated the use of the above methods for designing high strength concrete mixes with a number of examples in a separate monograph. The Indian Standard code stipulates that only controlled concrete should be used for prestressed concrete construction. The exact requirements of specifications with regard to the acceptance criteria for concrete generally vary from one code to the other. The British code CP 110 ~ 1972 stipulates that not more than 5 percent of the test results to fall below the 28-day characteristic cube strength, while the revised I. S. code-456 and I. S. code-1343 prescribes a similar stipulation of 5 percent. The corresponding requirement according to the American Concrete Institute standard ACI 214-65 is that not more than 10 percent of the test results should be below the specified design strength.

Strength Requirements

The minimum 28-day cube compressive strength prescribed in the I. S. code is 42 N/mm^2 for pre-tensioned members and 35 N/mm^2 for post-tensioned members. The ratio of standard cylinder to cube strength may be assumed to be 0.8 in the absence of any relevant test data. A minimum cement content of 350 and 400 kg/m^3 is prescribed for pre-tensioned and post-tensioned work respectively mainly to cater to the durability requirement. In high strength concrete mixes the water content should be as low as possible with due regard to adequate workability and the concrete should be suitable for compaction by the means available at site. It is the general practice to adopt vibration to achieve thorough compaction of concrete used for prestressed members. To safeguard against excessive shrinkage, the B. S. code prescribes that the cement content in the mix should preferably not exceed 530 kg/m^3. The specified works cube strength of 40 N/mm^2 required for prestressed members can be easily achieved even at the age of 7 days by using rapid hardening portland cement.

Deformation Characteristics of Concrete

The complete stress-strain characteristics of concrete in compression is not linear, but for loads not exceeding 30 percent of the crushing strength, the load deformation behaviour may be assumed to be linear. The deformation characteristics of concrete under short term and sustained loads is necessary for determining the flexural strength of beams and for evaluating the modulus of elasticity required for the computation of deflections of prestressed members.

The short term static modulus of elasticity specified in most of the codes corresponds to the secant modulus determined from an experimental stress-strain relation exhibited by standard specimens under loads of one-third of the cube compressive strength of concrete. The modulus of elasticity of concrete increases with the average compressive strength of concrete, but at a decreasing rate. Several empirical formulae have been recommended in various national codes for the computation of Young's modulus of elasticity of concrete, invariably expressed as a function of the compressive strength of concrete.

1. According to the Indian Standard code of practice

$$E_c/(\text{N} \cdot \text{mm}^{-2}) = 5\ 688(f_{cu})^{1/2} \qquad (8.1)$$

2. According to the provisions of the European Concrete Committee (C. E. B)

$$E_c/(\text{N} \cdot \text{mm}^{-2}) = 6\ 000(f_{cu})^{1/2} \qquad (8.2)$$

3. The American Concrete Institute (ACI 318-71) recom-mends a formula of the type

$$E_c/(\text{N} \cdot \text{mm}^{-2}) = 4\ 800(f_{cy})^{1/2} \qquad (8.3)$$

4. The British code for structural concrete specifies the values of the modulus of elasticity of concrete which is related to the cube strength as detailed in Tab. 8.1.

For light weight concrete having a density between 1 400 kg/mm^3 and 2 300 kg/mm^3, the values given in Tab. 8.1 should be multiplied by $(D_c/2\ 300)^2$, where is the density of the light weight aggregate concrete in kg/mm^3. It is also recommended that under conditions of sustained loading, appropriate allowances for shrinkage and creep are to be made.

5. Specifies the average values for the modulus of elasticity and Poisson's ratio of concrete as specified by the German specification, DIN-4227 which is given in Tab. 8.2.

Recent experimental investigations have indicated that the present Indian Standard code provisions generally overestimate the modulus of elasticity of concrete in the high strength range. A simple relationship of the type, $E_c = 750 f_{cu}$, is suggested for predicting the static secant modulus of elasticity of high strength concrete expressed in N/mm² in the range of 40 to 45 N/mm².

Tab. 8.1 Values of Modulus of Elasticity (CP110-1972)

Cube strength at the appropriate age or stage considered /(kN·mm⁻²)	Modulus of elasticity /(kN·mm⁻²)
20	25
25	26
30	28
40	31
50	34
60	36

Tab. 8.2 Modulus of Elasticity and Poisson's Ratio of Concrete (DIN-4227)

Concrete quality /(N·mm⁻²)	M-22.5	M-30	M-45	M-60
Modules of elasticity /(kN·mm⁻²) E_c	24	30	35	40
Poisson's ratio	0.15~0.18	0.17~0.20	0.20~0.25	0.25~0.30

Words and Expressions

impermeability [im͵pəːmiə'biləti]　　n. 不渗透性,防水性,气密性
abrasion resistance　　抗磨性
aggregate ['ægrigit]　　n. 骨料
slump [slʌmp]　　n. 坍塌,塌落
angularity [͵æŋgju'læriti]　　n. 棱角,有角性,斜度

monograph ['mɔnəgrɑːf]　n. 专(题)论(文)，专著论文单行本
stipulate ['stipjuleit]　v. 规定
secant ['siːkənt]　n. 正割，割线

Passage B　High Tensile Steel in Prestressed Concrete Members

Types of High Tensile Steel

For prestressed concrete members, the high strength steel used generally consists of wires, bars or strands. High carbon steel ingots are hot rolled into rods and cold drawn through a series of dies to reduce the diameter and increase the tensile strength. Cold drawn wires of 5~12 mm diameter are usually employed as individual wires or in parallel bundles or cables. The small diameter wires of 2~4 mm are mostly used in the from of strands comprising of two, three or seven wires. The helical form of twisted wires in the strand substantially improves the bond strength. High tensile steels usually contain 0.7~0.8 percent carbon, 0.6 percent manganese and about 0.1 percent silica. Bars are first hot rolled and subsequently heat treated.

The process of cold drawing through dies decrease the ductility of the wires. The cold drawn wires are subsequently tempered to improve its properties. The tempering or ageing or stress relieving of the wires at 150~420 ℃ results in the improvement of the tensile strength. The hard drawn steel wires which are indented or crimped are preferred for pre-tensioned work due to their superior bond characteristics.

The wires, which are used individually or in wire cables, are generally from 5~7 mm in diameter and have ultimate tensile strengths of about 1 500 N/mm^2. The strands, which are commonly used, vary in nominal diameter from 10~44 mm. The high tensile steel bars, which are commonly employed in prestressing, vary in diameter from 10 ~ 32 mm. The ultimate tensile strength of the bars does not vary appreciably with the diameter since the high strength of the bars is due to alloying rather than cold working as in the case of wires.

Strength Requirements

The ultimate tensile strength of high tensile steel varies with the diameter of the wire. The tensile strength is somewhat less for wires of larger diameter than those of smaller diameter. The minimum ultimate tensile strength requirements as per the Indian Standard code are as outlined in Tab. 8.3.

The high tensile alloy steel bars should have a minimum ultimate tensile strength of 1 000 N/mm^2. Unlike ordinary mild steel, high tensile wires have no well defined yield point and it is necessary to refer to the proof-stresses, which correspond to specified permanent strains.

Tab. 8.3 Characteristic Strength of High Tensile Steel (IS 1343)

(a) **High tensile wires**

Nominal diameter /mm	Characteristic strength/(N · mm^{-2})		
	Plain wires cold worked stress relieved	Indented wires	Plain wires as drawn
3	1 900	1 900	—
4	1 750	1 750	1 750
5	1 600	1 600	1 600
7	1 500	—	—
8	1 400	—	—

(b) High tensile bars

Nominal diameter/mm	Characteristic strength/($N \cdot mm^{-2}$)
10~40	1 000

(c) strands

Designation		Minimum breaking load/kN
2.0 mm	2 ply	13.00
2.9 mm	2 ply	26.00
6.3 mm	7 ply	45.36
7.9 mm	7 ply	70.31
9.5 mm	7 ply	95.25
11.1 mm	7 ply	127.00
12.5 mm	7 ply	167.83
15.2 mm	7 ply	231.33

The 0.2 percent proof stress for high tensile steel wires and alloy bars used for prestressed work should not be less than 80 percent of the minimum ultimate tensile strength. An important characteristic of the steel used in prestressing is the plasticity of the steel at stresses near the ultimate stress. This is essential to achieve progressive failure of the prestressed concrete members with sufficient warning before final failure. To avoid the possibility of brittle failure, the normal practice is to specify that the prestressing steel will have a minimum elongation at rupture. The Indian Standard code prescribes a minimum limit of 2 percent for the elongation at rupture, when tested over a gauge length of 200 mm.

Relaxation of Stress in Steel

When a high tensile steel wire is stretched and maintained at a constant strain, the initial force in the wire does not remain constant but decreases with time. The decrease of stress in steel at constant strain is termed as relaxation. In a prestressed member, the high tensile wires between the anchorages are more or less in a state of constant strain. However, the actual relaxation will be rather less than that indicated by a test of a wire at constant length, as there will be a shortening of the member due to other causes. With the high tensile steels at elevated stresses in excess of 0.01 percent proof stress, the relaxation of stress has been observed and its magnitude increases with the magnitude of initial stress. If the stress is maintained constantly, the material exhibits a plastic strain over and above the initial elastic strain, generally referred to as creep.

The cold drawn steels creep more than heat treated or tempered steels due to their lower magnitude of 0.01 percent proof stress. The phenomenon of creep is influenced by the chemical composition, micro structure, grain size and variables in the manufacturing process, which results in changes in the internal crystal structure. Several hypotheses for explaining the mechanism of creep in steel are presented by several investigators.

The steel in a prestressed concrete member strictly does not remain under a constant condition of either stress or strain. The most severe condition generally occurs at the stage of initial stressing; subsequently, the strain in the steel reduces as the concrete deforms under the prestressing force.

The code provisions for the relaxation of stress in steel is based on the results of the 1 000 hours relaxation test on the wires. Experience has shown that the loss recorded over a period of about

1 000 hours from an initial stress of 70 percent of the tensile strength is about the same as the loss experienced over a period of four years from an initial stress of 60 percent of the tensile strength. According to Stussi, the relaxation curves obtained over 1 000 hours can be extrapolated by a logarithmic plot. The Indian Standard specification I. S. 1785 prescribes the 1 000-hour relaxation test with a relaxation of stress not exceeding 70 N/mm^2, for cold drawn stress relieved wires. In the absence of this, the 100-hour relaxation test is also provided with a limiting value of relaxation stress of 46.7 N/mm^2. Experiments have shown that the reduction in relaxation stress is possible by preliminary over stressing. A preliminary over stress of 5 ~ 10 percent maintained for two or three minutes results in a considerable reduction in the magnitude of relaxation. Some of the codes permit temporarily over stressing with corresponding lower magnitudes of relaxation stress.

Stress Corrosion

The phenomenon of stress corrosion in steel is particularly dangerous since it results in sudden brittle fractures. Stress corrosion cracking results from the combined action of corrosion and static tensile stress, which may be either residual or externally applied. This type of attack in alloys is due to the internal metallurgical structure which is influenced by the composition, the heat treatment and the mechanical processing. The causes of the susceptibility of high tensile steels to stress corrosion are manifold. Schwier has reported that heat treated wires are specially prone to stress corrosion fractures when compared to drawn wires. If the ducts of post-tensioned members are not grouted, there is the possibility of stress corrosion leading to a catastrophic failure of the structure.

There are other common types of corrosion frequently encountered in prestressed concrete constructions such as pitting corrosion and chloride corrosion. A critical review of the different types of corrosion of high tensile steel in structural concrete is reported elsewhere. Some of the important protective measures to prevent stress corrosion include protection from chemical contamination, protective coatings for high tensile steel and grouting of ducts immediately after prestressing operations.

Hydrogen Embrittlement

Atomic hydrogen is liberated due to the action of acids on high tensile steels. This penetrates into the steel surface making it brittle and resulting in fractures on being subjected to tensile stress. Even small amounts of hydrogen are sufficient to cause considerable deterioration in the tensile strength of high tensile steel wires.

Use of high alumina cement, blast furnace slag cement which is rich in sulphides when used to make prestressed concrete are likely to give rise to hydrogen embrittlement. Use of dissimilar metals such as aluminium and zine for sheaths to house, high tensile steel wires also results in hydrogen embrittlement. Minute traces of sulphur which come in contact with high tensile steel wires in the presence of moisture results in drastic reduction in the strength due to hydrogen embrittlement.

In order to prevent hydrogen embrittlement, it is essential that steel is properly protected from the action of acids. Protective coverings like bituminous crepe paper covering during transport reduces the chances of contamination. The wires should be protected from rain and excessive humidity by storing them in dry conditions.

Cover Requirements for Tendons

The alkaline environment of portland cement concrete generally protects embeded tendons and other reinforcements against corrosion from various environmental agencies. However, the carbonation of hydrated cement gradually progresses from the surface to the interior of concrete, thus reducing the effective protection provided by the concrete to prevent the rusting of steel tendons. Many codes have provided for minimum cover requirement in this regard. It is pertinent to note that not only the thickness of cover but also the density of concrete in the cover are important to provide effective protection to steel.

The Indian Standard code (I. S-1343) provides with a minimum clear cover of 20 mm for protected pre-tensioned members, while it is 30 mm or the size of the cable (whichever is bigger) in the case of protected post-tensioned members. If the prestressed members are exposed to the atmosphere, these cover requirements are increased by 12 mm.

The British code (CP-110) recommendations regarding cover are somewhat more comprehensive since the prescribed values of cover are related to the severity of the environmental conditions and the quality of concrete, as indicated by the cement content and the water/cement ratio. Four different degrees of exposure from mild to very severe are identified and the nominal covers recommended are compiled in Tab. 8.4, in which, degrees of exposure are

Mild: completely protected against weather or aggressive conditions except for exposure to normal weather conditions briefly during construction.

Moderate: Sheltered from severe rain and against freezing while saturated with water-buried concrete and concrete under water.

Severe: Exposed to driving rain, alternate wetting, drying and freezing while wet-subject to heavy condensation or corrosive fumes.

Very Severe: Exposed to sea water or moorland water with

abrasion.

Tab. 8.4 British Cover Code Requirements for Tendons and Reinforcement in Prestressed Concrete

Condition of exposure	Minimum cement content* /(kg·mm^{-2})	Maximum free water/cement ratio**	Nominal cover for concrete of /mm		
			Grade 30	Grade 40	Grade 50 and over
Mild	300	0.65	15	15	15
Moderate	319	0.55	30	25	20
Severe	370	0.45	40	30	25
Very severe	—	—	—	60	50
Subject to deicing salts	310	0.55	50+	40+	25

* For maximum size of aggregate of 10 mm

** Where water/cement ratio can be strictly controlled

\+ For concrete with entrained air only

Words and Expressions

strand [strænd]   n. 绞合金属绳，钢绞线
ingot ['iŋgət]   n. 金属锭，锭(块，坯，料)
helical ['helikl]   n. &a. 螺旋(的)，螺线(的)
manganese [ˌmæŋgə'niːz]   n. 锰
silica ['silikə]   n. 硅石，二氧化硅
ductility [dʌk'tiləti]   n. 延性
indent [in'dent]   v. 刻痕
alloy ['ælɔi]   n. 熔成合金，生产合金
elongation [ˌiːlɔŋ'geiʃən]   n. 拉长，伸长，延伸率
extrapolate [ek'stræpəleit]   v. 推断，推知

logarithmic [ˌlɔgə'riðmik] *a.* 对数的
residual [ri'zidjuəl] *a.* 残余的
metallurgical [ˌmetə'lə:dʒikəl] *a.* 冶金的
susceptibility [səˌseptə'biləti] *n.* 敏感度
grout ['graut] *v.* 灌浆
deterioration [diˌtiəriə'reiʃən] *n.* 恶化，降低
alumina [ə'lju:minə] *n.* 矾土
hydrogen embrittlement （钢的）氢脆
sheath [ʃi:θ] *n.* 外包层
bituminous [bi'tju:minəs] *a.* （含）沥青的
alkaline ['ælkəlain] *n. ;a.* 碱性(的)
portland cement 普通硅酸盐水泥
tendon ['tendən] *n.* 钢筋束
hydrate ['haidreit] *v.* 水合，水化

Dialogue

A: Volume 8 in the Tender Document is the Bidding Document. You should complete and submit it according to the requirement.

B: Yes, I have noticed that. We should include in it a Letter of Agreement, a Construction Method Statement, Quality and Safety Programs and a Bill of Quantities.

A: These are extremely important for you to win the project.

B: Though our company is an experienced one, we still hope you can explain these key points in more details.

A: All right. The Letter of Agreement is the promise of the Contractor to the Employer. It gives the total initial contract price, the commencement date, the completion date and the maintenance period. This letter should be signed by the legal representative of the Contractor.

B: What about other important points such as the Advanced Payment,

Liquidated Damages, Bank Guarantee...

A: There is an attachment of the Letter of Agreement. You should put in your opinions in the form attached.

Words and Expressions

Construction Method Statement　施工方案
Quality and Safety Program　质量和安全措施
Bill of Quantity　工程量清单
Commencement Date　开工日期
initial contract price　初始合同价
advanced payment　预付款
liquidated damages　罚金
bank guarantee　银行担保
attachment [ə'tætʃmənt] n. 附件
legal representative　法人代表

Unit 9

Passage A Strength of Reinforced Concrete

9.1 Strength of Concrete

Normal concrete consists of coarse and fine aggregate, cement, and water. The materials are mixed together until a cement paste is developed, filling most of the voids in the aggregates and producing a uniform dense concrete. The plastic concrete is then placed in a mold and left to set, harden, and develop adequate strength.

The strength of concrete depends upon many factors and may vary within wide limits with the same production method. The main factors that affect the strength of concrete will be described below.

Water-cement Ratio

The water-cement ratio is the most important factor affecting strength of concrete. For compete hydration of a given amount of cement, a water-cement ratio (by weight) equal to 0.25 is needed. A water-cement ratio more than 0.1 higher is needed for the concrete to be reasonably workable. This means that a water-cement ratio of more than 0.35 by weight must be chosen. This ratio corresponds to 4 gallons of water per sack of cement (94 lb) (or 17.5 L per 50 kg of cement). Good workability is attained when the ratio exceeds 0.5. The relation between water-cement ratio and compressive and flexural strengths of normal weight concrete is shown in Tab. 9.1.

Tab. 9.1 Typical Relation Between Water-cement Ratio and Compressive and Flexural Strengths of Normal Weight Concrete

Water-cement ratio			Probable strength of concrete at 28 days			
By weight	Gallons per sack (94 lb)	Liters per sack (50 kg)	Compressive		Flexural	
			psi	N/mm^2	psi	N/mm^2
0.35	4.0	17.5	6 300	41	650	4.5
0.40	4.5	20.0	5 800	40	610	4.2
0.44	5.0	22.0	5 400	37	590	4.1
0.49	5.5	24.5	4 800	33	560	3.9
0.53	6.0	26.5	4 500	31	540	3.7
0.58	6.5	29.0	3 900	27	500	3.5
0.62	7.0	31.0	3 700	25	490	3.4
0.67	7.5	33.5	3 200	22	450	3.1
0.71	8.0	35.5	2 900	20	430	3.0

The Properties and Proportions of Concrete Constituents

Concrete is a mixture of cement, aggregate, and water. An increase in the cement content in the mix and the use of well-graded aggregate increase the strength of concrete.

The Method of Mixing, Placing, Degree of Compaction and Curing

The use of mechanical concrete mixers and proper time of mixing have favorable effects on strength of concrete. In addition, the use of vibrators produces dense concrete with minimum percentage of voids. A void ratio of 5 percent may reduce the concrete strength by about 30 percent.

The curing conditions exercise an important influence on the strength of concrete. Both moisture and temperature have a direct effect on the hydration of cement. The longer the period of moist storage, the greater the strength. If the curing temperature is higher than the initial temperature of casting, the resulting 28-day strength of concrete is reached earlier than 28 days.

The Age of Concrete

The strength of concrete increases appreciably with age, and hydration of cement continues for months. In practice, the strength of concrete is determined from cylinders or cubes tested at age of 7 days and 28 days. As a practical assumption, concrete at 28 days is 1.5 times as strong as at 7 days; the range varies between 1.3 and 1.7. The British code of practice accepts concrete if its strength at 7 days is not less than two-thirds of the required 28-day strength. The strength of concrete at 28 days can be calculated by an empirical formula relating this strength with that at 7 days as follows

$$f'_{c(28)}/(\text{psi}) = f'_{c(7)} + 30\sqrt{f'_{c(7)}} \qquad (9.1)$$

where $f'_{c(28)}$ and $f'_{c(7)}$ are ultimate strengths at 28 and 7 days, respectively, in pounds per square inch. In S.I. units,

$$f'_{c(28)}/(\text{N} \cdot \text{mm}^{-2}) = f'_{c(7)} + 2.5\sqrt{f'_{c(7)}} \qquad (9.2)$$

$$f'_{c(28)}/(\text{kgf} \cdot \text{cm}^{-2}) = 8\sqrt{f'_{c(7)}} \qquad (9.3)$$

For a normal portland cement, the increase of strength with time, relative to 28-day strength, is as follows:

Age	7 days	14 days	28 days	3 months	6 months	1 year	2 years	5 years
strength ratio	0.67	0.86	1.0	1.17	1.23	1.27	1.31	1.35

Loading Conditions

The compressive strength of concrete is estimated by testing a cylinder or cube to failure in a few minutes. Under sustained loads for years, the ultimate strength of concrete is reduced by about 30 percent. Under 1-day sustained loading, concrete may lose about 10 percent of its compressive strength. Sustained loads, as well as dynamic and impact effects if they occur on the structure, should be considered in the design of reinforced concrete members.

The Shape and Dimensions of the Tested Specimen

The common sizes of concrete specimens used to predict the compressive strength are either 6 by 12 in. (150 by 300 mm) cylinders or 6 in. (150 mm) cubes. When a given concrete is tested in compression by means of cylinders of like shape but of different sizes, the larger specimens give lower strength indexes. Tab. 9.2 gives the relative strength, for various sizes of cylinders, as a percentage of the strength of the standard cylinder; the heights of all cylinders are twice as the diameters.

Sometimes concrete cylinders of nonstandard shape are tested. The greater the ratio of specimen height to diameter, the lower the strength indicated by the compression test. To compute the equivalent strength of the standard shape, the results must be multiplied by a correction factor. Approximate values of the correction factor are given in Tab. 9.3, extracted from ASTM C42-57 The relative strengths of a cylinder and a cube for different compressive strengths are shown in Tab. 9.4.

Tab. 9.2 Effect of Size of Compression Specimen on Strength of Concrete

Size of cylinder /inch	/mm	relative compressive strength
2×4	50×100	1.09
3×6	75×150	1.06
6×12	150×300	1.00
8×16	200×400	0.96
12×24	300×600	0.91
18×36	450×900	0.86
24×48	600×1 200	0.84
36×72	900×1 800	0.82

Tab. 9.3 Strength Correction Factor for Cylinders of Different Height-diameter Ratios

Ratio	2.0	1.75	1.50	1.25	1.10	1.00	0.75	0.5
Strength correction factor	1.00	0.98	0.96	0.94	0.90	0.85	0.70	0.50
Strength relative to standard cylinder	1.00	1.02	1.02	1.04	1.11	1.11	1.43	2.00

Tab. 9.4 Relative Strengths of Cylinder vs. Cube

Compressive/psi	1 000	2 200	2 900	3 500	3 800	4 900	5 300	5 900	6 400	7 300
strength/($N \cdot mm^{-2}$)	7.0	15.5	20.0	24.5	27.0	24.5	37.0	41.5	45.0	51.5
Strength ratio of cylinder to cube	0.77	0.76	0.81	0.87	0.91	0.93	0.94	0.95	0.96	0.96

9.2 Required Strengths of Concrete

Different strengths of a particular concrete may be required in the design of reinforced concrete structures; these include the compressive, tensile, flexural, and shear strengths. Although sometimes it is difficult to perform tests to predict these strength, it is possible to relate them to the compressive strength.

Compressive Strength

In designing structural members, it is assumed that the concrete resists compressive stresses and not tensile stresses; therefore, compressive strength is the criterion of quality of concrete. The other concrete stresses can be taken as a percentage of the compressive strength, which can be easily and accurately determined from tests. Specimens used to determine compressive strength may be cylindrical, cubical, or prismatic.

Test specimens in form of a cylinder 6 in. (150 mm) across and 12 in. (300 mm) high are usually used to determine the strength of concrete.

Cube specimens with sides of 6 in. (150 mm) or 8 in. (200 mm) are used in great Britain, Germany, and other parts of Europe.

Prism specimens are used in France. the USSR, and other countries and are usually 70 by 70 by 350 mm or 100 by 100 by 500 mm. They are cast with longer sides horizontal and tested, like cubes, in a position normal to the position of cast.

Before testing, the specimens are moist-cured and then tested at the age of 28 days by gradually applying a static load until rupture occurs. The rupture of the concrete specimen may be caused by the applied tensile stress (failure in cohesion), the applied shearing stress (sliding failure), the compressive stress (crushing failure), or combinations of the above stresses.

The failure of the concrete specimen can be in one of three modes. First, under axial compression. the concrete specimen may fail in shear. Resistance to failure is due to both cohesion and internal friction. The shear resistance may be represented by Coulomb's equation and Mohr's circle.

The second type of failure results in the separation of the specimen into columnar pieces by what is known as splitting or columnar fracture. This failure occurs when the strength of concrete is high, and lateral expansion at the end bearing surfaces is relatively unrestrained.

The third type of failure is seen when a combination of shear and splitting failure occurs.

Tensile strength

Concrete is a brittle material, and it cannot resist the high tensile stresses that are important when considering cracking, shear, and torsional problems. The low tensile capacity can be attributed to the high stress concentrations in concrete under load, so that a very high stress is reached in some portions of the specimen. causing microscopic cracks, while the other parts of the specimen are subjected to low stress.

Direct tension tests are not reliable for predicting the tensile strength of concrete, due to minor misalignment and stress concentrations in the gripping devices. An indirect tension test in the form of splitting a 6 by 12 in . (150 by 300 mm) cylinder was suggested by the Brazilian Fernando Carneiro. The test is usually called the splitting test, but it was also named the Brazilian test (although it was developed independently in Japan). In this test, the concrete cylinder is placed with its axis horizontal in a compression testing machine. The load is applied uniformly along two apposite lines on the surface of the cylinder through two plywood pads.

Considering an element on the vertical diameter and at a distance y from the top fibers, the element is subjected to a compressive stress

$$f_c = \frac{2P}{\pi LD}\left(\frac{D^2}{y(D-Y)} - 1\right) \qquad (9.4)$$

and a tensile stress

$$f'_{sp} = \frac{2P}{\pi LD} \qquad (9.5)$$

where

P——the compressive load on the cylinder

D and L——the diameter and length of the cylinder

For a 6 by 12 in. (150 by 300 mm) cylinder and at a distance $y = D/2$, the compression strength is

$$f_c = 0.026\ 5P \qquad (9.6)$$

and the tensile strength

$$f'_{sp} = 0.008\ 8P = f_c/3 \qquad (9.7)$$

Cubes can also be subjected to a splitting test. The load is applied through semicylindrical steel rollers resting against the cube.

The splitting strength f'_{sp} can be related to the compressive strength of concrete in that it varies between 6 and 7 times $\sqrt{f'_c}$ for normal concrete and between 4 and 5 times $\sqrt{f'_c}$ for lightweight concrete. The direct tensile stress f'_t can also be estimated from the split test; its value varies between $0.5 f'_{sp}$ and $0.7 f'_{sp}$. The smaller of the above values applies to higher strength concrete. The splitting strength f'_{sp} can be estimated as 10 percent of the compressive strength up to $f'_c = 6\ 000$ psi (42 N/mm^2). For higher values of compressive strength, f'_{sp} can be taken as 9 percent of f'_c. In general, the tensile strength of concrete ranges from 7 to 11 percent of its compressive strength with an average of 10 percent. The lower the compressive strength, the higher the relative tensile strength.

Flexural Strength (Modulus of Rupture)

Experiments on concrete beams have shown that ultimate tensile strength in bending is greater than the tensile stresses obtained by direct or splitting tests.

Flexural strength is expressed in terms of the modulus of rupture of concrete (f_r), which is the maximum tensile stress in concrete in bending. The modulus of rupture can be calculated from the flexural formula used for elastic materials, $f_r = (M \cdot c)/I$, by testing a plain concrete beam. The beam of 6 by 6 by 28 in. (150 by 150 by 700) is supported on a 24-in. (600 mm) span and loaded to rupture by two loads, 4 in. (100 mm) on either side of the center. A smaller beam of by 4 by 4 by 20 in. (100 by 100 by 500 mm) on a 16-in. (400 mm) span may also be used.

The modulus of rupture of concrete ranges between 11 and 23 percent of the compressive strength. A value of 15 percent can be assumed for strengths of about 4 000 psi (28 N/mm^2).

A number of empirical formulae relate the modulus of rupture of concrete to its compressive strength. The CEB gives the value of the modulus of rupture f_r as

$$f_r = 9.5 \sqrt{f'_c} \text{ psi} = 0.79 \sqrt{f'_c} \text{ N/mm}^2 \qquad (9.8)$$

The ACI Code (318-83) prescribes the value

$$f_r = 7.5 \sqrt{f'_c} \text{ psi} = 0.62 \sqrt{f'_c} \text{ N/mm}^2 \qquad (9.9)$$

The modulus of rupture as related to strength obtained from the split test on cylinders may be taken as

$$f_r = (1.25 \text{ to } 1.50) f'_{sp} \qquad (9.10)$$

Shear Strength Pure shear is seldom encountered in reinforced concrete members, as it is usually accompanied by the action of normal forces. An element subjected to pure shear breaks transversely into two parts. Therefore, the concrete element must be strong enough to resist the applied shear forces.

Shear strength may be considered as 20 to 30 percent greater than the tensile strength of concrete, or about 12 percent of its compressive strength. The ACI Code allows an ultimate shear strength of $2\sqrt{f'_c}$ psi ($0.17\sqrt{f'_c}$ N/mm^2) on plain concrete sections.

Words and Expressions

hydration [hai'dreiʃən] n. 水化作用
vibrator ['vai'breitə] n. 振捣器
rupture ['rʌptʃə] n. 破裂,裂开,折断,挠曲
cohesion [kəu'hiːʒən] n. 结合力
torsional ['tɔːʃənəl] a. 扭转的
plywood ['plaiwud] n. 胶合板
modulus ['mɔdjuləs] n. 模数,指数
transversely ['trænsvəːsli] adv. 横向地

Passage B Probability of Failure

To initiate the discussion of probability of failure let us consider a special case. A single structural component (see Fig. 9.1) is acted upon by a random load. The mean and standard deviation of the load are assumed known and denoted as $\overline{P}$ and σ_p, respectively. The load induces a stress, and if the beam cross-sectional area is considered to be deterministic, the stress is a random variable S with mean $\overline{S} = \overline{P}/A$ and standard deviation. Assume that the stress-strain properties of the material are as shown in Fig. 9.1. The yield stress of the material is considered to be a random variable R with known mean and standard deviation (denoted as $\overline{R}$ and σ_R). If failure is defined to exist when the load-induced stress equals or exceeds the material yield stress, then failure exists when

$$S \geqslant R \qquad (9.11)$$

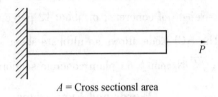

A = Cross sectionsl area

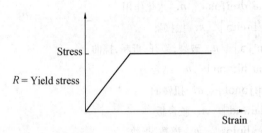

Fig. 9.1 Hypothetical example

If one defines a new random variable F as

$$F \equiv R - S \tag{9.12}$$

failure, therefore, occurs when F is less than or equal to zero.

Consider first the special case where both R and S have independent normal PDF's. That is,

$$p(s) = \frac{1}{\sigma_S \sqrt{2\pi}} \exp\left\{ -\frac{1}{2} \left(\frac{S - \bar{S}}{\sigma_S} \right)^2 \right\} \tag{9.13}$$

and

$$p(r) = \frac{1}{\sigma_R} \sqrt{2\pi} \exp\left\{ -\frac{1}{2} \left(\frac{r - \bar{R}}{\sigma_R} \right)^2 \right\} \tag{9.14}$$

The random variable F has normal probability density function because it is a linear combination of two normally distributed random variables. It directly follows that

$$\bar{F} = \bar{R} - \bar{S} \tag{9.15}$$

and

$$\sigma_F^2 = \sigma_R^2 + \sigma_S^2 \qquad (9.16)$$

Therefore, the PDF of F is

$$P(f) = \frac{1}{\sigma_F \sqrt{2\pi}} \exp\left\{ -\frac{1}{2}\left(\frac{f - \overline{F}}{\sigma_F}\right)^2 \right\} \qquad (9.17)$$

and is shown in Fig. 9.2. Since failure occurs when F is less than or equal to zero, the probability of failure P_f is

$$P_f = Pr[F \leq 0] = \int_{-\infty}^{0} p(f) \, df \qquad (9.18)$$

Fig. 9.3 shows the calculated probability of failure for several combinations of coefficients of variations of R and S (i.e. $\rho_R = (\sigma_R/\overline{R})$ and $\rho_S = (\sigma_S/\overline{S})$) and several central safety factor values where

$$\text{central safety factor} \equiv C_0 \equiv \frac{\overline{R}}{\overline{S}} \qquad (9.19)$$

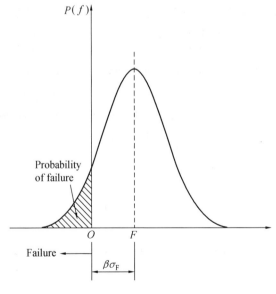

Fig. 9.2 Failure probability

It is apparent from Fig. 9.2 that $F=0$ occurs at $\overline{F}$ minus standard deviations. Therefore, the value is directly related to the probability of failure. It follows that knowing $\bar{F}$ and σ_F and setting $F = 0$, i. e.,

$$F = 0 = \bar{F} - \beta\sigma_F$$

results in a value of equal to

$$\beta = \frac{\bar{R} - \bar{S}}{\sqrt{\sigma_R^2 + \sigma_S^2}} = \frac{\bar{F}}{\sigma_F} \qquad (9.20)$$

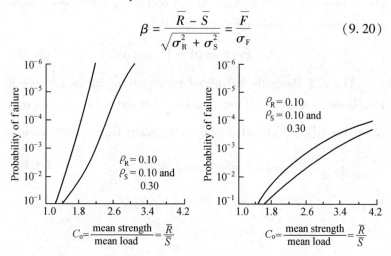

Fig. 9.3 **Probability of failure for normal R and S**

Where, β is called the reliability index or safety index. Equation (9.17) can be transformed into the standardized normal PDF

$$p(u) = \frac{1}{\sqrt{2\pi}}\exp\left\{-\frac{1}{2}u^2\right\} \qquad (9.21)$$

if

$$U \equiv \frac{F - \bar{F}}{\sigma_F} \qquad (9.22)$$

The probability of failure is now obtained when $F \leq 0$ (i. e., $u = -(\bar{F}/\sigma_F) = -\beta$), and is

$$P_f = Pr[F \leq 0] = \int_{-\infty}^{-\beta} p(u)\,du \qquad (9.23)$$

Consider the second special case, where R and S have

independent log-normal PDF's. Taking the natural logarithm of the ratio of R divided by S, it follows that

$$\ln\left(\frac{R}{S}\right) = \ln R - \ln S \qquad (9.24)$$

If one define the following transformations,

$$X \equiv \ln R$$
$$Y \equiv \ln S$$
$$Z \equiv \ln\left(\frac{R}{S}\right)$$

then Eq. (9.24) becomes

$$Z = X - Y \qquad (9.25)$$

where X, Y, and Z are all normally distributed. It follows directly from Eq. (9.25) that

$$\overline{Z} = \overline{X} - \overline{Y} \qquad (9.26)$$

and

$$\sigma_Z^2 = \sigma_X^2 + \sigma_Y^2 \qquad (9.27)$$

and

$$p(z) = \frac{1}{\sigma_Z \sqrt{2\pi}} \exp\left\{-\frac{1}{2}\left(\frac{z - \overline{Z}}{\sigma_Z}\right)^2\right\} \qquad (9.28)$$

It now remains that we find $\overline{X}, \overline{Y}, \sigma_X^2$, and σ_Y^2 in terms of $\overline{R}$, $\overline{S}$, σ_R^2 and σ_S^2.

The mean and variance of the transformed variable (i.e., X) can be expressed in terms of the mean and variance of the log-normal random variable.

It follows from the power series expansion of $\ln(1+\varepsilon)$ that this is equal to ε for small values of ε. In the present case, ε is equal to the coefficient of variation squared; therefore, this approximation is acceptable. Thus, it is possible to use the following approximate relationships

$$\bar{X} = \ln \bar{R} \qquad (9.29)$$

and

$$\sigma_X^2 = \rho_R^2 \qquad (9.30)$$

where ρ_R = coefficient of variation of R. Therefore, Eq. (9.26) and (9.27) become

$$\bar{Z} = \ln \bar{R} - \ln \bar{S} \qquad (9.31)$$

and

$$\sigma_R^2 = \rho_R^2 + \rho_S^2 \qquad (9.32)$$

Recall that failure occurs when $R-S < 0$, or, alternatively, when $\dfrac{R}{S} < 1$. Taking the natural logarithm of both sides, we obtain $\ln(\dfrac{R}{S}) = \ln R - \ln S \leqslant 0$. Therefore, failure exists when Z as defined by Eq. (9.25) is equal to or less than zero.

The normal PDF. in Eq. (9.28) can be transformed into the standardized normal PDF by defining

$$U^* \equiv \frac{Z - \bar{Z}}{\sigma_Z} \qquad (9.33)$$

It directly follows that

$$p(u^*) = \frac{1}{\sqrt{2\pi}} \exp\left\{-\frac{1}{2}(u^*)^2\right\} \qquad (9.34)$$

The probability of failure obtained when $Z \leqslant 0$ (i. e., $u^* = -(\bar{Z}/\sigma_Z)$), and is

$$P_f = \Pr[Z \leqslant 0] = \int_{-\infty}^{-\beta^*} p(\mu^*)\,d\mu^* \qquad (9.35)$$

where

$$\beta^* = \frac{\ln \bar{R} - \ln \bar{S}}{\sqrt{\rho_R^2 + \rho_S^2}} \qquad (9.36)$$

Eq. (9.23) and (9.35) are very similar in form. In the former equation, R and S are assumed to be normally distributed and is evaluated by using Eq. (9.20). The latter equation is for log-normally distributed R and S random variables, and u^* is evaluated by using Eq. (9.36). In both cases.

Words and Expressions

sectional ['sekʃənl] *a.* 截面的
stress-strain 应力-应变
PDF 概率密度函数
specification [ˌspesifi'keiʃən] *n.* 规定,规范,技术要求
serviceability ['səːvisəbiliti] *n.* 使用(中)的可靠性(或舒适程度)

Dialogue

A: I'm sorry to say I'm not satisfied with your speed. You should have achieved more progress.
B: It is because of the bad weather. We can speed up when the weather permits. Still we think it is too tight to reach the schedule. ...
A: But you have to reach it. The Construction Period for the project is ten months. There is no room for discussion for the Key Dates fixed in the Tender Document.
B: We hope the section schedules can be rearranged.
A: Your rearrangement must not affect the general schedule. If your schedule can't match the Key Dates, it will be a big problem for the erection contractor.
B: We know that. We'll submit our modified section schedules to you for approval.

Words and Expressions

general schedule 总进度表
construction schedule 施工进度表

section schedule　分期进度表
construction period　工期
Key Dates　关键日期
erection contractor　安装承包商

Unit 10

Passage A Steel Structures

10.1 Types of Structural Steel Members

The function of a structure is the principal factor determining the structural configuration. Using the structural configuration along with the design loads, individual components are selected to properly support and transmit loads throughout the structure. Steel members are selected from the rolled shapes adopted by the American Institute of Steel Construction (AISC) (also given by American Society for Testing and Materials (ASTM) A6 Specification). Of course, welding permits combining plates and/or other rolled shapes to obtain any shape the designer may require.

Typical rolled shapes, the dimensions for which are found in the AISC Manual are shown in Fig. 10.1. The most commonly used section is the wide-flange shape (Fig. 10.1(a)) which is formed by hot rolling in the steel mill. The wide-flange shape is designated by the nominal depth and the weight per foot, such as a W18×97 which is nominally 18 in. deep (actual depth = 18.95 in. according to

AISC Manual) and weighs 97 pounds per foot. (In SI units the W18× 97 section could be designated W460×142, meaning nominally 460 mm deep and having a mass of 142 kg/m). Two sets of dimension are found in the AISC Manual, one set stated in decimals for the designer to use in computations, and another expressed in fractions (1/16 in. as the smallest increment) for the designer to use on plans and shop drawings. Rolled W shapes are also by ANSI/ASTM A6 in accordance with web thickness as Groups I through V, with the thinnest web sections in Group I.

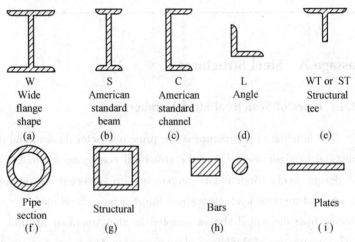

Fig. 10.1 Standard rolled shapes

The American Standard beam (Fig. 10.1(b)), commonly called the I-beam, has relatively narrow and sloping flanges and a thick web compared with the wide-flange shape. Use of most I-beams has become relatively uncommon because of excessive material in the web and relative lack of lateral stiffness due to the narrow flanges.

The channel (Fig. 10.1(c)) and angle (Fig. 10.1(d)) are commonly used either alone or in combination with other sections. The channel is designated, for example, as C12×20.7, a nominal 12-in.

deep channel having a weight of 20.7 pounds per foot. Angles are designated by their leg length (long leg first) and thickness, such as, $L6 \times 4 \times \frac{3}{8}$.

The structural tee (Fig. 10.1(e)) is made by cutting wide-flange or I-beams in half and is commonly used for chord members in trusses. The tee is designated, for example, as WT5×44, where the 5 is the nominal depth and 44 is the weight in pounds per foot; this tee is cut from a W10×88.

Pipe sections (Fig. 10.1(f)) are designated "standard", "extra strong", and "double-extra strong" in accordance with the thickness and are also nominally prescribed by diameter; thus 10-in. diam double-extra strong is an example of a particular pipe size.

Structure tubing (Fig. 10.1(g)) is used where pleasing architectural appearance is desired with exposed steel. Tubing is designated by outside dimensions and thickness, such as, $8 \times 6 \times \frac{1}{4}$.

The sections shown in Fig. 10.1 are all hot-rolled; that is, they are formed from hot billet steel (blocks of steel) by passing through rolls for numerous times to obtain the final shapes.

Many other shapes are cold-formed from plate material having a thickness not exceeding 1 in., as shown in Fig. 10.2.

Regarding size and designation of cold-formed steel members, there are no truly standard shapes even though the properties of many common shapes are given in the cold-formed Steel Design Manual. Various manufacturers produce many proprietary shapes.

Tension Members

The tension member occurs commonly as a chord member in a truss, as diagonal bracing in many types of structures, as direct support for balconies, as cables in suspended roof systems, and as

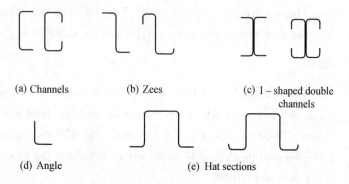

Fig. 10. 2   Some cold-formed shapes

suspension bridge main cables and suspenders that support the roadway. Typical cross-sections of tension members are shown in Fig. 10.3.

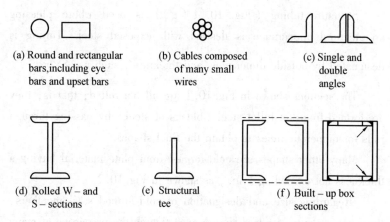

Fig. 10. 3   Typical tension members

Compression Members

Since compression member strength is the function of the cross-sectional shape (radius of gyration), the area is generally spread out as much as is practical. Chord members in trusses, and many interior columns in buildings are examples of members subject to axial compression. Even under the most ideal condition, pure axial compression is not attainable; so design for "axial" loading assumes that the effect of any small simultaneous bending may be neglected. Typical cross sections of compression members are shown in Fig. 10.4.

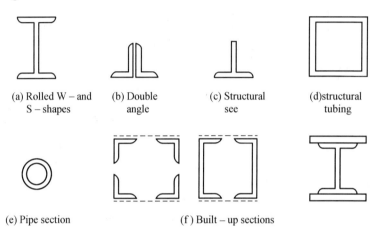

Fig. 10.4 **Typical compression members**

Beams

Beams are members subjected to transverse loading and are most efficient when their area is distributed to be located at the greatest practical distance from the neutral axis. The most common beam sections are the wide-flange (W) and I-beams (S) (Fig. 10.5(a)), as well as smaller rolled I-shaped sections designated as "miscellaneous shapes" (M).

For deeper and thinner-webbed sections than can economically be rolled, welded I-shaped sections (Fig. 10.5(b)) are used, including stiffened plate girder. For moderate spans carrying light loads, open-web "joists" are often used (Fig. 10.5(c)). These are parallel chord truss-type members used for the support of floors and roofs. The steel may be hot-rolled or cold-formed.

For beams (known as lintels) carrying loads across window and door opening, angles are frequently used; and for beams (known as girts) in wall panels, channels are frequently used.

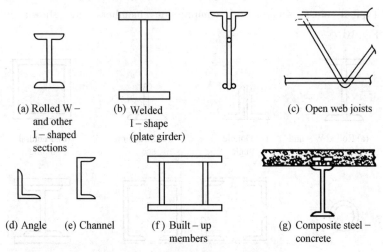

(a) Rolled W− and other I−shaped sections

(b) Welded I−shape (plate girder)

(c) Open web joists

(d) Angle (e) Channel (f) Built−up members

(g) Composite steel − concrete

Fig. 10.5　Typical beam members

Bending and Axial Load

When simultaneous action of tension or compression along with bending occurs, a combined stress problem arises and the type of member used will be dependent on the type of stress that predominates. A member subjected to axial compression and bending is usually referred to as a beam-column. The aforementioned illustration of types of members resisting various kinds of stress is intended only to show common and representative types of members and not to be all inclusive.

10.2　Steel Structures

Structures may be divided into three general categories (a) framed structures, where elements may consist of tension members, columns, beams and members under combined bending and axial load; (b) shell-type structures, where axial stresses predominate; and (c) suspension-type structures, where axial tension predominates the principal support system.

Framed Structures

Most typical building construction is in this category. The multistory building usually consists of beams and columns, either rigidly connected or having simple end connections along with diagonal bracing to provide stability. Even though a multistory building is three-dimensional, it usually is designed to be much stiffer in one direction than the other; thus it may reasonably be treated as a series of plane frames. However, if the framing is such that the behavior of the members in one plane substantially influences the behavior in another plane, the frame must be treated as three-dimensional space frame.

Industrial buildings and special one-story building such as churches, schools, and arenas, generally are either wholly or partly framed structures. Particularly the roof system may be a series of plane trusses, a space truss, a dome, or it may be part of a flat or gabled one-story rigid frame.

Bridge are mostly framed structures, such as beams and plate girders, or trusses, usually continuous.

Most of this text is devoted to behavior and design of elements in framed structures.

Shell-type Structures

In this type of structure the shell serves a use function in addition to participation in carrying loads. One common type where the main stress is tension is the containment vessel used to store liquids (for both high and low temperatures), of which the elevated water tank is a notable example. Storage, bins, tanks, and the hulls of ships are other examples. On many shell-type structure, a framed structure may be used in conjunction with the shell.

On walls and flat roofs the "skin" elements may be in compression while they act together with a framework. The aircraft body is another example.

Shell-type structures are usually designed by a specialist and are not within the scope of this text.

Suspension-type Structures

In the suspension-type structure tension, cables are major supporting elements. A roof may be cable-supported. Probably the most common structure of this type is the suspension bridge. Usually a subsystem of the structure consists of a framed structure, as in the stiffening truss for the suspension bridge. Since the tension element is the most efficient way of carrying load, structures utilizing this concept are increasingly used.

Many unusual structures utilizing various combinations of framed, shell-type, and suspension-type structures have been built. However, the typical designer must principally understand the design and behavior of framed structures.

Words and Expressions

increment ['inkrimənt]   *n.* 增量,增值
excessive [ik'sesiv]   *a.* 过多的,极度的
billet ['bilit]   *n.* 钢坯
diagonal bracing   对角支撑
radius of gyration   回转半径
miscellaneous [ˌmisi'leinjəs]   *a.* 混杂的
predominate [pri'dɔmineit]   *v.* 控制,支配
suspension-type   悬索型
multistory ['mʌltiˌstɔːri]   *n.* 多层
gable ['ɡeibl]   *n.* 山墙(gabled   *a.* 有山墙的)

shell-type structure   壳型结构

Passage B   Probabilities of Occurrence of Tornado Winds

Consider an area A_0, say, a one-degree longitude-latitude square, and let the tornado frequency in that area (i.e., the average number of tornado occurrences per year) be denoted by $\bar{n}$. The probability that a tornado will strike a particular location during one year is assumed to be

$$P(S) = \bar{n}\frac{\bar{a}}{A_0} \qquad (10.1)$$

where $\bar{a}$ is the average individual tornado area. In certain applications, for example, the design of nuclear power plants, rather than the probability $P(S)$, it is of interest to estimate the probability $P(S, V_0)$ that a tornado with maximum wind speeds higher than some specified value V_0 will strike a location in any one year. This probability can be written as

$$P(S, V_0) = P(V_0)P(S) \qquad (10.2)$$

where $P(V_0)$ is the probability that the maximum wind speed in a tornado will be higher than V_0.

Probabilities $P(S)$ in the United States were estimated, based on Eq. 1 in which $\bar{n}$ was estimated from 13-year frequency data, $\bar{a} = 2.82$ sq. miles and $A_0 = 4\,780 \cos \varphi$, where φ is the latitude at the center of the one-degree square considered. Estimated probabilities $P(V_0)$ are shown in the map of the United States. These estimates are based upon observations of 1612 tornadoes during 1971 and 1972. It is noted that in estimating the probabilities, it was assumed that tornado path areas are the same throughout the contiguous United States.

The maximum speed of the tornado corresponding to a specified probability of occurrence can be estimated. According to Technical

Basis for Interim Regional Tornado Criteria (WASH – 1300 (UC – 11)), "In order to adequately protect public health and safety, the determination of the design basis tornado is based on the premise that the probability of occurrence of a tornado that exceeds the Design Basis Tornado (DBT) should be on the order of 10^{-7} per nuclear power plant". The required probability $P(V_0)$ is then determined from the relation

$$P(V_0)P(S) = 10^{-7} \qquad (10.3)$$

where the value of $P(S)$ for the location considered is taken from the above estimated probabilities. The wind speed corresponding to the probability $P(V_0)$ so determined can be then obtained. The average tornado intensity with a 10^{-7} probability per year for each 5-degree square in the contiguous United States, based on Eq. 10.3 and Eq. 10.1 and Eq. 10.2, is obtained.

For nuclear power plant design purposes, the contiguous United States are divided into three tornado intensity regions. The corresponding tornado winds are given in Tab. 10.1.

Tab. 10.1 Regional Tornado Winds

Region	Maximum Speed V_{max} /mph	Rotational Speed V_{rot} /mph	Translational Speed V_{tr} /mph	Radius of Maximum Rotational Wind speed R_m /ft
I	360	290	70	150
II	300	240	60	150
III	240	190	50	150

The pressure drop due to the passage of tornadoes can be estimated from the equation for the cyclostrophic wind. Using the relation = dr/dt, Eq. 10.2 can be written as

$$\frac{dp}{dt} = \frac{V_{tr}}{R_m}\rho V_t^2 \qquad (10.4)$$

where p is the pressure, t is the time, V_{tr} is the translational speed, ρ is the air density, R_m is the radius of maximum rotational wind speed, and V_{tr} is the maximum tangential wind speed. Assuming R_m is typically 150 ft for intense tornadoes and $V_t \approx V_{rot}$, Eq. 10.4, in which the parameters of Tab. 10.1 are used, yields approximately the values of Tab. 10.2. Using as a point of departure tornado risk maps, a regionalization of tornado risks which divides the contiguous United States into four areas was proposed. Regional tornado occurrence rate (per mi^2 per year) were estimated from a 29-year (1950 ~ 1978) data bank maintained by the National Severe Storms Forecast Center comprising about 20 000 reported tornadoes. These regional occurrence rates are corrected to account for:

Tab. 10.2 Regional Pressure Drops and Pressure Drop Rate

Region	Total Pressure Drop /psi	Rate of Pressure Drop /(psi · sec^{-1})
I	3.0	2.0
II	2.25	1.2
III	1.5	0.6

1. Failure to record tornado intensity, which affects about 10% of the total number of reported tornadoes. This correction is based on the assumption that unrated tornadoes may be apportioned among the various intensity categories according to the reported tornado frequencies for those categories.

2. Temporal variations in tornado reporting efficiency. The number of reported annual tornado occurrences in the United States has increased from about 250 in 1950 to 850 in 1979. The growing trend in the number of reported tornadoes during this period has been

ascribed to a corresponding increase in population density. An explicit relation to this effect has been proposed. Corrections accounting for tornado reporting efficiencies were effected by averaging the 1971 ~ 1978, 1970 ~ 1978, 1969 ~ 1978, and 1950 ~ 1978 data assuming that the true occurrence rates are equal to the largest of these estimations.

3. Possible errors in the rating of tornado intensities on the basis of observed damage. The reason for the occurrence of such errors is that maximum tornado winds in practice are not measured, but inferred, largely on the basis of professional judgment, from observations of damage to buildings, signs, and so forth.

4. Inhomogeneous distribution along the tornado path of buildings and various other objects susceptible of being damaged. In the possible absence of such objects over the portions of the tornado path where the winds are the highest (or even over the entire tornado path), the rating of the tornado is bound to be in error. The effect of corrections for such errors is to increase the estimated probability of occurrence of tornadoes with higher intensities.

5. Variation of tornado intensity along the tornado path. Accounting to this factor results in smaller estimated risks of high tornado winds than would be the case if the maximum tornado winds (by which tornado intensities are rated) were uniform along the entire path. Corrections effected, based upon the analysis of documented tornadoes, led to risk reductions by a factor of about five for F4 tornadoes and about ten for F6 tornadoes.

Words and Expressions

tornado [tɔː'neidəu] n. 龙卷风
contiguous [kən'tigjuəs] a. 相邻的
premise ['premis] n. 前提

tangential [tæn'dʒenʃəl] *a.* 切线的,切向的
temporal ['tempərəl] *a.* 暂时的
inhomogeneous [ˌinhəumə'dʒiːniəs] *a.* 不同类的
cyclostrophic wind   旋衡风

Dialogue

A: According to the General Construction Schedule, you should have finished the installation of the HVAC (Heating, Ventilation and Air-Conditioning) works by now.

B: It was held up because of the delay of your approval of equipment procurement.

A: I can't agree with you there. You proposed to use different equipment instead of that described in the specification. So we needed to investigate if the proposed equipment was suitable or not.

B: But the investigation lasted too long.

A: I think we'd better stop arguing about that. The important thing to do now is to finish it in the shortest possible time.

B: We can finish it within 30 days.

A: Can you try to do it within 25 days?

B: We'll try our best.

Words and Expressions

General Construction Schedule   施工总进度计划
equipment procurement   设备采购

Unit 11

Passage A Selecting a Framing Scheme

Nowhere in engineering design is intuition more appropriate, more essential, than in preliminary design. There are no formulae that lead inexorably to the most appropriate framing scheme. Rather, a brilliant scheme for an unusual problem and even good workmanlike schemes for routine problems will be the result of heavy intuitive thinking based on years of experience. Engineers do well in cultivating and respect insights to structural behavior that cannot be proved easily or quickly by normal means.

Engineers should strive for redundancy in preparing a preliminary design. Redundancy means the ability of a structural frame to carry load by more than one path. In a redundant structure, if a beam would fail, the load supported by the beam could be carried to the ground by other means, such as catenary action of slab reinforcement. Redundant flexural members also tent to reduce deflection. It is not necessary that the secondary members carry load without distress—only that total collapse be avoided.

It is more difficult, but not impossible, to provide alternate load paths for columns and tension members or hangers. For such members, redundancy is defined as limiting failure to the immediate area surrounding the failed member. That is, beams and slabs can have sufficient continuity and anchorage at each end to prevent complete collapse if support is withdrawn at one end, if a column fails, the surrounding floors and roofs might sag considerably but should not endanger life by dropping to the ground.

Engineers should favor the selection of structural frames with a high degree of redundancy because structural redundancy is essential in preventing progressive collapse. Some owners require their engineers to design the structure to minimize the risk of progressive collapse. If a member or a frame cannot be made redundant, more than normal care should be taken in its design. In addition, conservative engineers will use a slightly higher factor of safety for such members.

Selection of the best, most appropriate framing scheme will normally proceed in the following manner.

Find a Typical Bay or Panel

A typical bay is the one most often repeated in the building with the fewest modifications. A bay is the space from one column line to the next, extending across the entire width of the building, and a panel is the space between four columns (Fig. 11. 1). It will be advantageous to consider a bay rather than a panel if the column spacing has not yet been established or if the number and spacing of columns can be improved. If a typical bay does not include the most critical members, they should be considered in determining the size of corresponding members in the typical bay.

If there are no typical bays, it may mean that the structural frame (and also probably the building) will be very expensive to design and build. The engineer's first task then is to rearrange building elements so as to achieve some degree of regularity and uniformity to produce a typical bay. This need not result in compromises to the architectural design or use of the building. Nor need it result in a monotonous building facade.

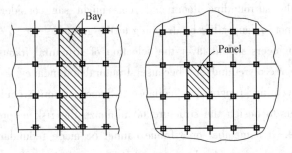

Fig. 11.1 Typical bays and panels

Determine Minimum Slab Thickness

Embedment of electrical conduit (used in almost all building slabs) requires a slab thickness of at least 3 in. in pan slabs. Other slabs should be at least 4 in. thick. If the slab has under-floor electrical distribution, greater thickness will be required. Minimum thickness may also be determined by requirements for fire protection.

The building code governing design and construction of the structural frame will specify the required fire resistance rating in hours. Some owners might want to increase the code-required fire rating, or the engineer may recommend a higher rating in some case. Frames for small, low buildings frequently do not need a fire rating. The floors for larger and taller buildings must have a 2-hr or, sometimes, 3-hr fire rating. Columns and walls frequently must have a 2 to 4-hr fire rating.

The fire resistance rating of a structure is the time a structure is exposed to heat of a fire during which (a) it can support itself and superimposed loads without collapse and (b) the temperature rise on the unexposed surface of the concrete does not exceed 250 °F as an average for the entire surface, nor 325 °F at any one point. The purpose of the second criterion is to guard against ignition of combustible material in contact with the unexposed surface. fire ratings are determined by testing a structural member or assembly in a furnace exposed to heat with standard time-temperature curve.

Because concrete is more resistant to fire than steel is, a concrete frame will continue to carry load at least as long as the stress in reinforcing steel does not exceed the yield point of the steel. Neglecting load factors and capacity reduction factors (this is permissible when considering fire rating), the stress in steel at service loads will generally be less than 0.6 f_y, where f_y is the yield strength at room temperature. The yield strength of hot-rolled structural grade steel will fall below 0.6 f_y, when its temperature rises somewhat higher than 1 000 °F.

If sufficient concrete cover is provided to insulate the steel so that the steel temperature remains below 1 000 °F, the structure will continue to carry design service loads. In solid slabs, the required cover is 3/4 in. for 1.5 hr resistance rating, 1.25 in. for 3-hr fire resistance rating, and 1.5 in. for 4-hr fire resistance rating. A little more cover is required for joists because the bottom steel is exposed on three sides with minimum concrete cover. A little less cover is needed when light-weight concrete is used due to its better insulating properties.

The portion of a structure exposed to fire in a limited area is stronger than a isolated, simple span structure because thermal expansion is resisted by the unexposed portions of the structure and because negative moment regions of flexural members are better protected from heat than mid-span positive moment regions. For these reasons, it is generally considered that concrete cover required by the ACI Code (Code 7.7.1) is adequate for the fire resistance rating normally required.

The important parameters in determining whether a frame will meet the second requirement limiting heat transmission are the thickness of concrete slabs or walls and the type of aggregate used. Aggregate will usually be selected on the basis of local availability, cost, or other reasons. Thus the required fire resistance rating will determine the minimum thickness of slabs in the structural frame.

The CRSI Committee on Fire Ratings describes standard fire test procedures, tabulates the results of many fire test, and describes analytical methods for computing fire resistance ratings of reinforced concrete floor members and assemblies.

Determine Column Spacing

Column locations may be given by the architect or owner and cannot be changed. If so, proceed to the next step. When the engineer participates in or influences the decision on column locations, building use as well as structural economy must be considered. Column spacing closer than about 15 ft will not normally be economical because floor slabs spanning 15 ft will probably have minimum, or nearly minimum, thickness and reinforcement. Shorter spans do not decrease the cost of slabs as fast as the cost of columns increases. Likewise, column spacing more than about 35 to 40 ft increase the cost substantially. The cost of longer floor spans rises faster than the reduction in cost of larger but fewer columns. Because the total weight of a given building is nearly constant, the total required cross-sectional area of columns in one story is nearly constant. Considering only the columns, fewer and larger columns minimize formwork and reduce the cost.

Columns should be located in walls where they interfere the least with use of the building. In the exterior wall, the spacing and shape of columns affects tenant's views of outside scenery. Columns projecting into the room should be coordinated with interior space planning.

In most buildings there is a limited choice in column spacing. For example, in a 78-ft-wide building, columns can be spaced 2 at 39 ft, 3 at 26 ft, 4 at 19.5 ft, 5 at 15.6 ft, or with minor variations of unequal spans. Considering the use of the building, only one or two of these options will be practical.

Sketch as Many Different Framing Schemes as Can be Conceived for the Typical Bay

The following observations will assist an engineer in arriving at appropriate framing schemes.

A. Solid slabs are economical up to about 20 ft and span up to about 25 ft.

B. Pan slabs are economical up to about 35 or 40 ft. Where possible, pan slabs should span the long direction of a panel, and beams the short direction, so that the depth of beams and joists can be equal for simplicity of formwork.

C. For spans over 35 to 40 ft, prestressing or deep beams must be used to control deflections. Prestressing may unduly restrict the owner from drilling holes through the slab to provide electrical, computer, or telecommunication services at new locations when office layouts change. Deep beams may interfere with mechanical service below the floor slab and raise the cost unduly.

D. Building for residential occupancy, such as hotels, motels, dormitories, and apartments frequently use an exposed solid concrete slab as the finished ceiling. Surface preparation of the concrete and panting is less expensive than a separate hung ceiling. In residential buildings, spans are within the economic limits of one- or two-way solid slabs.

E. Hospitals also frequently have short spans, but mechanical services may prevent using the slab as an exposed ceiling.

F. Office buildings, banks, commercial structures, and schools generally require spans in the range of 25 to 40 ft.

G. Office buildings and, sometimes, other buildings require flexibility in providing electrical, computer, and telecommunication services at new locations when wall or room layouts change. These services can be provided by drilling. into under-floor ducts an inch or two below top of the slab or by a "poke-thru" system. Electrical under-floor ducts can be installed easily in solid slabs and accommodated in pan slabs by using a pan size 2 in. shallower, in Fig. 11.2. In a poke-thru system, holes are drilled through the slab to reach conduit just below the floor slab. Poke-thru holes in pan slabs will usually miss reinforcement. In solid slabs, poke-thru holes will sometimes cut reinforcement, and design of slabs should make allowance for this possibility.

H. Slabs exposed to the weather, especially freezing weather when deicing salts may be used (as in a parking garage), should have 2 in. cover to reinforcement, a water/cement (W/C) ratio of 0.40 or less, and should slope at least 1/4 in. per foot to drains. A slope of 3/8 to 1/2 in. is better. Thin slabs (as in a pan slab) should be avoided as their long-term durability may not be satisfactory.

I. Flat slabs with drop panels and column capitals are generally the most economical system to support heavy live loads exceeding about 150 to 200 psf.

J. Members with a minimum size controlled by reasons other than stress should be used so that they are fully stressed. Some common examples follow:

a. If slabs must have a minimum thickness for fire protection, the structural system should use slabs that span far enough to stress minimum reinforcement in slabs to the limit. For example, a 4.5-in. slab can span at least 7 or 8 ft.

b. Minimum reinforcement in flexural members should be fully stressed under load.

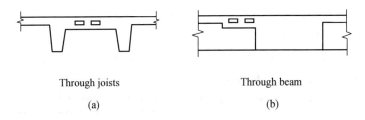

Through joists Through beam

(a) (b)

Fig. 11.2 Sections through joints and under-floor electrical ducts

c. Concrete walls used for partitions should also be used to carry vertical loads and resist lateral loads.

d. The spacing of lightly loaded columns of minimum size should be increased so that they are loaded to capacity.

K. Use members for more than one purpose, if possible, for example:

a. When nonstructural, noncombustible floor finishes such as terazzo or ceramic tile are used, they may be included in the thickness required for fire protection.

b. Walls and long, narrow columns can carry vertical loads, resist lateral loads, and serve as partitions. Basement walls can serve as grade beams to simplify foundation construction. Other walls can serve as girders to transfer column loads or other heavy loads to support points.

c. Use temperature reinforcement in a one-way solid slab as

flexural reinforcement in the long direction of a two-way slab (if the aspect ratio of the slab is two or less).

L. For economy of structural frame, consider framing system approximately in the following order:

   a. Flat plate without drop panels or column capitals.
   b. Flat plate with wide shallow beams.
   c. Pan slabs.
   d. Flat slab with drop panels and column capitals.
   e. Waffle slabs.
   f. Slabs and deep beams.

Tab. 11.1 summarizes some of the advantages and disadvantages of common framing schemes:

Tab. 11.1 Advantages and Disadvantages of Framing Schemes

Framing Scheme	Advantages	Disadvantages
Flat plate	Least cost for work	Low capacity in shear
	Exposed ceilings	excessive concrete
	Minimum thickness	for longer span
	Fast erection	
	Flexible column location	
Flat slab	Economical for heavy loading	Expensive formwork
Flat banded slab	Extends span range of flat plate	Formwork costs more than for flat plate
	Minimum thickness	unless many reuses of flying form are possible
Pan slabs and wide beams	Minimum concrete and steel, minimum weight, hence, reduces size of columns and	Unattractive for a ceiling formwork costs more than for flat plate

Continued table

Framing Scheme	Advantages	Disadvantages
	foundations	
	Long span in one direction, Easy poke-thru electrical	
Skip joists and wide beams	Similar to pan slabs but use slightly less concrete	Similar to pan slabs Joists must be designed as beams
	Can place mechanical equipment in space between joists	
Dome slab	Attractive exposed ceiling	Formwork costs more than for pan slab
		Uses more concrete than a pan slab
Deep one-way beams and one-way	Long span in one direction	Beams inerfere with mechanical work
		Expensive formwork
Deep two-way beams and two-way	Long span in two direction	Beams interfere with mechanical work
	Small deflection	Expensive formwork

As indicated above, it is necessary to consider the use to which a building will be put in preparing a preliminary design. The number, size, and location of holes through the floor system for stairs, elevators, and mechanical shafts, as well as the building mechanical system, will affect the selection of a structural frame. Under-floor electrical ducts may be installed. Laboratory equipment may be unusually sensitive to vibrations and deflections. Heavy static or moving loads should be considered in the preliminary design.

Design Each Proposed Framing Scheme for the Typical Bay

Each design should be complete, establishing all final concrete outlines and quantities of concrete and reinforcing steel. Loads should be as exact as possible, but approximations may be used when laborious computations are needed to determine exact loading. For example, a series of concentrated loads may be approximated as a uniform load. In establishing member outlines, make allowance for nontypical panels that are more highly stressed than the typical panel. However, design of the typical panel should be for loads and conditions of the typical panel.

Review Proposed Framing Schemes

To help ensure that all criteria have been considered and that the best schemes have been delineated, the engineer should review the preliminary designs. The review will be focused on important issues if the engineer asks pertinent questions about each proposed scheme. For example, the engineer should consider the following:

A. How can I make this scheme more repetitive, more economical?

B. What compromises on the part of the owner, architect, mechanical engineer, or other people will make this scheme better?

C. What compromises on my part will make this scheme more acceptable to other people?

D. What feature of this scheme costs the most? Can it be modified to reduce the cost?

E. What feature of this scheme makes it least acceptable to other people? Can that feature be modified to make it more acceptable?

Considering all schemes, the engineer should ask the following

A. Have I considered all possible scheme?

B. Is there a better way to meet design objectives?

Words and Expressions

intuition [ˌintju'iʃən] n. 直觉
inexorably [in'eksərəbli] adj. 不屈不挠地
workmanlike ['wəːkmənlaik] a. 精巧的
cultivate ['kʌltiveit] v. 启发
redundancy [ri'dʌndənsi] n. 超静定的,徐徐的
catenary [kə'tiːnəri] a. 悬链的
distress [di'stres] n. 损坏
sag [sæɡ] v. 下垂
conservative [kən'səːvətiv] a. 保守的
monotonous [mə'nɔtənəs] a. 单调的
facade [fə'sɑːd] n. 立面
conduit ['kɔndit] n. 导管
ignition [iɡ'niʃən] n. 点燃
combustible [kəm'bʌstəbl] a. 可燃的
solid slab n. 实心板
joist [dʒɔist] n. 肋
isolated ['aisəleitid] a. 独立的(孤立的)
unduly [ˌʌn'djuːli] adv. 过度地,不适当地
terrazzo [te'rɑːtsəu] n. 水磨石
ceramic [si'ræmik] a. 陶瓷的
aspect ratio n. 长宽比

Passage B High-rise Buildings

Introduction

It is difficult to define a high-rise building. One may say that a low-rise building ranges from 1 to 2 stories. A medium-rise building probably ranges between 3 or 4 stories up to 10 or 20 stories or more.

Although the basic principles of vertical and horizontal subsystem design remain the same for low-, medium-, or high-rise buildings, when a building gets high the vertical subsystems become a controlling problem for two reasons. Higher vertical loads will require larger columns, walls, and shafts. But, more significantly, the overturning moment and the shear deflections produced by lateral forces are much larger and must be carefully provided for.

The vertical subsystems in a high-rise building transmit accumulated gravity load from story to story, thus requiring larger column or wall sections to support such loading. In addition these same vertical subsystems must transmit lateral loads, such as wind or seismic loads, to the foundations. However, in contrast to vertical load, lateral load effects on buildings are not linear and increase rapidly with increase in height. For example under wind load, the overturning moment at the base of buildings varies approximately as the square of a buildings may vary as the fourth power of buildings height, other things being equal. Earthquake produces an even more pronounced effect.

When the structure for a low-or medium-rise building is designed for dead and live load, it is almost an inherent property that the columns, walls, and stair or elevator shafts can carry most of the horizontal forces. The problem is primarily one of shear resistance. Moderate addition bracing for rigid frames in "short" buildings can easily be provided by filling certain panels (or even all panels) without increasing the sizes of the columns and girders otherwise required for vertical loads.

Unfortunately, this is not is for high-rise buildings because the problem is primarily resistance to moment and deflection rather than shear alone. Special structural arrangements will often have to be made and additional structural material is always required for the columns, girders, walls, and slabs in order to make high-rise buildings sufficiently resistant to much higher lateral deformations.

As previously mentioned, the quantity of structural material required per square foot of floor of a high-rise building is in excess of that required for low-rise buildings. The vertical components carrying the gravity load, such as walls, columns, and shafts, will need to be strengthened over the full height of the buildings. But quantity of material required for resisting lateral forces is even more significant.

With reinforced concrete, the quantity of material also increases as the number of stories increases. But here it should be noted that the increase in the weight of material added for gravity load is much more sizable than steel, whereas for wind load the increase for lateral force resistance is not that much more since the weight of a concrete buildings helps to resist overturn . Additional mass in the upper floors will give rise to a greater overall lateral force under the seismic effects.

In the case of either concrete or steel design, there are certain basic principles for providing additional resistance to lateral to lateral forces and deflections in high-rise buildings without too much sacrifice in economy.

1. Increase the effective width of the moment-resisting subsystems. This is very useful because increasing the width will cut down the overturn force directly and will reduce deflection by the third power of the width increase, other things remaining constant. However, this does require that vertical components of the widened subsystem be suitably connected to actually gain this benefit.

2. Design subsystems such that the components are made to interact in the most efficient manner. For example, use truss systems with chords and diagonals efficiently stressed, place reinforcing for

walls at critical locations, and optimize stiffness ratios for rigid frames.

3. Increase the material in the most effective resisting components. For example, materials added in the lower floors to the flanges of columns and connecting girders will directly decrease the overall deflection and increase the moment resistance without contributing mass in the upper floors where the earthquake problem is aggravated.

4. Arrange to have the greater part of vertical loads be carried directly on the primary moment-resisting components. This will help stabilize the buildings against tensile overturning forces by precompressing the major overturn-resisting components.

5. The local shear in each story can be best resisted by strategic placement if solid walls or the use of diagonal members in a vertical subsystem. Resisting these shears solely by vertical members in bending is usually less economical, since achieving sufficient bending resistance in the columns and connecting girders will require more material and construction energy than using walls or diagonal members.

6. Sufficient horizontal diaphragm action should be provided floor. This will help to bring the various resisting elements to work together instead of separately.

7. Create mega-frames by joining large vertical and horizontal components such as two or more elevator shafts at multistory intervals with a heavy floor subsystem, or by use of very deep girder trusses.

Remember that all high-rise buildings are essentially vertical cantilevers which are supported at the ground. When the above principles are judiciously applied, structurally desirable schemes can be obtained by walls, cores, rigid frames, tubular construction, and other vertical subsystems to achieve horizontal strength and rigidity.

Some of these applications will now be described in subsequent sections in the following.

Shear-wall Systems

When shear walls are compatible with other functional requirements, they can be economically utilized to resist lateral forces in high-rise buildings. For example, apartment buildings naturally require many separation walls. When some of these are designed to be solid, they can act as shear walls to resist lateral forces and to carry the vertical load as well. For buildings up to some 20 storise, the use of shear walls is common. If given sufficient length, such walls can economically resist lateral forces up to 30 to 40 stories or more.

However, shear walls can resist lateral load only the plane of the walls (i. e. not in a direction perpendicular to them). There fore, it is always necessary to provide shear walls in two perpendicular directions can be at least in sufficient orientation so that lateral force in any direction can be resisted. In addition, that wall layout should reflect consideration of any tensional effect.

In design progress, two or more shear walls can be connected to from L-shaped or channel-shaped subsystems. Indeed, internal shear walls can be connected to from a rectangular shaft that will resist lateral forces very efficiently. If all external shear walls are continuously connected, then the whole buildings acts as tube, and connected, then the whole buildings acts as a tube, and is excellent Shear-Wall Systems resisting lateral loads and torsion .

Whereas concrete shear walls are generally of solid type with openings when necessary, steel shear walls are usually made of trusses. These trusses can have single diagonals, " X " diagonals, or " K " arrangements. A trussed wall will have its members act essentially in direct tension or compression under the action of view, and they offer some opportunity and deflection-limitation point of view, and they offer some opportunity for penetration between members. Of course, the inclined members of trusses must be suitable placed so as not to interfere with requirements for windows and for circulation service penetrations though these walls.

As stated above, the walls of elevator, staircase, and utility shafts form natural tubes and are commonly employed to resist both vertical and lateral forces. Since these shafts are normally rectangular or circular in cross-section, they can offer an efficient means for resisting moments and shear in all directions due to tube structural action. But a problem in the design of these shafts is provided sufficient strength around door openings and other penetrations through these elements. For reinforced concrete construction, special steel reinforcements are placed around such opening. In steel construction, heavier and more rigid connections are required to resist racking at the openings.

In many high-rise buildings, a combination of walls and shafts can offer excellent resistance to lateral forces when they are suitably located ant connected to one another. It is also desirable that the stiffness offered these subsystems be more-or-less symmetrical in all directions.

Rigid-frame Systems

In the design of architectural buildings, rigid-frame systems for resisting vertical and lateral loads have long been accepted as an important and standard means for designing building. They are employed for low-and medium means for designing buildings. They are employed for low-and medium up to high-rise building perhaps 70 or 100 stories high. When compared to shear-wall systems, these rigid frames both within and at the outside of a buildings. They also make use of the stiffness in beams and columns that are required for the buildings in any case, but the columns are made stronger when rigidly connected to resist the lateral as well as vertical forces though frame bending .

Frequently, rigid frames will not be as stiff as shear-wall construction, and therefore may produce excessive deflections for the more slender high-rise buildings designs. But because of this flexibility, they are often considered as being more ductile and thus less susceptible to catastrophic earthquake failure when compared with (some) shear-wall designs. For example, if over stressing occurs at certain portions of a steel rigid frame (i. e. , near the joint), ductility will allow the structure as a whole to deflect a little more, but it will by no means collapse even under a much larger force than expected on the structure . For this reason, rigid-frame construction is considered by some to be a "best" seismic-resisting type for high-rise steel buildings. On the other hand, it is also unlikely that a well-designed share-wall system would collapse.

In the case of concrete rigid frames, there is a divergence of opinion. It true that if a concrete rigid frame is designed in the conventional manner, without special care to produce higher ductility, it will not be able to withstand a catastrophic earthquake that can produce forces several times larger than the code design earthquake forces . Therefore, some believe that it may not have additional capacity possessed by steel rigid frames. But modern research and experience has indicated that concrete frames can be designed to be ductile, when sufficient stirrups and joinery reinforcement are designed in to the frame. Modern buildings codes have specifications for the so-called ductile concrete frames. However, at present, these codes often require excessive reinforcement at certain points in the frame so as to cause congestion and result in construction difficulties.

Even so, concrete frame design can be both effective and economical.

Of course, it is also possible to combine rigid-frame construction with shear-wall systems in one buildings, For example, the buildings geometry may be such that rigid frames can be used in one direction while shear walls may be used in the other direction.

Words and Expression

joint [dʒɔint]   n.   接头,节点
frame [freim]   n.   框架
earthquake forces   地震力
low-rise buildings   低层建筑
high-rise building   高层建筑
shear-wall systems   剪力墙结构
rigid-frame systems   框架结构

Dialogue

A: The quality of underground works is very critical to the whole project. You cannot expect a stable superstructure without a solid foundation.

B: That's true. We should ensure that the quality of the underground works in compliance with the specification.

A: Before going to the site, let's first look at the Execution Tracing Files for the underground works.

B: The files are all here, including the Execution Tracing Files, the Work Procedures and the Method Description. They record the drain works.

A: What have you done to ensure this part of works to be in compliance with the specification?

B: As you know, before the works started we submitted the Works Procedures for the execution of the works. And we followed the

Procedures strictly in our working.

A: What equipment do you use?

B: Well, we use a hydraulic machine. Before we began to insert the drains, we measured and marked the position for each drain, and then the hydraulic machine was employed for inserting. (on the construction site)

A: Good. You have done an excellent job, and I can see the drains are inserted vertically and they go to the exact length.

Words and Expressions

underground works 基础工程
critical ['kritikl] *a.* 至关重要的,关键的
superstructure ['sjuːpəˌstrʌktʃə] 上部结构
in compliance with 符合,与……一致
Execution Tracing Files 施工跟踪文件
drain works 排水工程
Work Procedures 施工程序文件
Method Description 施工方案
hydraulic machine 液压机
insert [in'səːt] *v.* 插入,打入
vertically [vəːtikəli] *adv.* 垂直地

Unit 12

Passage A Determining Concrete Outlines

Because one objective of preliminary design is to determine concrete outlines, design procedures leading directly to member sizes should be used. Computation related to reinforcement can be disregarded unless the quantity of steel is needed for cost estimates.

In the following discussion, experienced engineers will recognize appropriate exceptions.

The shortest and easiest design procedures should be used so that the engineer does not lose sight of objectives in the minutiae of detailed design. If the first preliminary design reveals potentially critical members or situations, a more accurate design procedure can be used on the second iteration.

An approximate moment analysis for gravity loads using moment factors will normally suffice even for spans falling a little outside ACI code limitations. Engineers can draw on their experience to use factors sufficiently accurate. Likewise, an approximate frame analysis by the Portal Method for lateral loads should be used. For tall or important buildings, a second iteration using more accurate methods of frame analysis may be necessary.

Do not vary the member size to fit the span and load except by increments of at least 30% to 50% or more. For example, if a 12-in. column is inadequate, increase the size to 16 to 18 in.. Make the member large enough for the most critical condition and vary the amount of reinforcement on the final design for less critical conditions.

Repeat bay size, panel size, story heights, and member size as often as possible. If, by inspection or by calculation, the load on a column or beam is more than 60% to 80% but less than 130% to 150% of the load on a similar member, use the same concrete outline for both members.

Do not use the maximum capacity of a concrete outline on the preliminary design. Unexpectedly higher loads, conditions slightly changed for the worse, and higher stresses by more precise analysis than those given by approximate analysis should all be accommodated by the initial concrete size. Only gross errors or significant changes clearly requiring a revised preliminary design should lead to changes in concrete outline on the final design. For large projects, a second iteration of preliminary design may be necessary to reduce the margin of uncertainty in preliminary sizes.

Because additional cement or other expensive ingredients are required to increase the strength of concrete, specify the lowest satisfactory concrete strength needed for the project. Normally a strength of $f'_c = 3$ ksi throughout the project will be satisfactory unless there is a specific reason to use higher strength concrete. Common reasons for using higher strength concrete are as follows:

1. Columns supporting more than three or four stories may be more economical if higher strength concrete is used in the lower stories.

2. If concrete strength in the column is more than 40% higher

than concrete strength in the floor, or $1.4 * 3 = 4.2$ ksi, increasing the strength of concrete on the floor may be the most economical solution to passing column loads through the floor.

3. Concrete exposed to freezing and thawing should have $f'_c =$ 4 ksi or more in addition to air entertainment to ensure durability.

4. Buildings more than 10 or 15 stories tall will usually be built on a fast schedule, requiring design strength in less than 28 days, typically 5 to 10 days. In such case, the 28 day strength will be higher. For example, construction loads may require 3 ksi strength at 7 days, but strength at 28 days will be 3.5 to 4 ksi. If permanent service loading is heavy than construction loading, the higher concrete strength may be used in design for service loading.

5. Floor framing usually does not require a concrete strength f'_c greater than 3 ksi because deflection usually control the selection of concrete outlines rather than concrete strength. Increasing the strength of concrete is an inefficient method for reducing deflection or increasing shear strength.

When foundations are expensive (as in deep foundations) or the building is tall, keep member sizes as small as possible reduce dead load. In other buildings, it is more economical to be a little generous with member sizes. When it is desirable to reduce dead load, lightweight concrete should be considered. Its higher cost may be offset by saving in columns and foundations. When slab thickness is controlled by the fire resistance rating required by the building code, the use of lightweight concrete will justify thinner slabs.

Individual Member Sizes

Considerations in establishing sizes of beams, slabs, and columns are discussed below.

Beams The size of beams is usually controlled by flexural compression at the point of maximum negative moment. Positive moment is less critical for beams because a flange is usually present to help resist flexural compression and because positive moment is usually smaller than negative moment.

The depth of the beam is usually established by other considerations, such as minimizing floor construction thickness, simplifying formwork, or minimizing deflection. In some cases, beam width may be established to fit within a narrow space or to allow the passage of pipes or ducts through the beam. After either width or depth is established, the other dimension is easily computed.

Use of compression steel should be avoided on preliminary design and should be used only on final design where necessary for strength to maintain established concrete outlines or to reduce long-term deflection.

For maximum economy, it is usually best to keep the percentage of reinforcement low so that reinforcing bars can be placed easily, especially at intersecting beams and columns. With a low percentage of reinforcement, bars can be widely spaced so that concrete can be placed easily around the bars.

Shear generally does not control the size of beams except in very short, heavily loaded members. Therefore, shear need not be computed for preliminary design.

The depth of long-span, shallow beams, especially those with little or no continuity, may be controlled by deflection. In such cases, deflection should be computed and member size revised if necessary. Often, deflection computations for the typical or most critical beam will also suffice for final computations for the entire structure.

One-way Slabs Thickness of solid slabs may be controlled by fire resistance requirements. Alternatively, the thickness will be controlled by ACI Code requirements to limit deflection. In some cases, thicker slabs may be used to limit deflection still further.

Shear and flexural stresses rarely, if ever, control the thickness of solid one-way slabs.

Two-way Slabs Minimum thickness of two-way slabs is $Ln/32.7$, where Ln is the clear span and steel $fy = 60$ ksi is used. Punching shear around columns from gravity loads and from unbalanced moment transferred by shear will further control the design. Either a thicker slab, larger column, column capitals, or shear-head reinforcement may be necessary. Engineers should be especially conservative in selecting proportions to meet shear requirements if enlargement of columns or column capitals will be difficult during final design.

Finally, deflection behavior of the slab should be considered even though the minimum thickness is exceeded. The appropriate time to compute deflection is during preliminary design rather than during final design when changes to correct unsatisfactory behavior may be difficult.

Decreases more rapidly away from the support than shear does.

Pan Slabs Shear and negative moment will control the width, spacing and depth of joists. Positive moment will rarely, if ever, control concrete outlines of pan slabs. Increasing the width of joists near the support by using tapered pans is more effective in resisting negative moment than in resisting shear. Moment d Deflection behavior of joists on long spans and joists with little continuity should be checked during preliminary design. If computed deflection is excessive, deeper or wider joists can be used. Such changes are more difficult to make during final design when the engineer is restricted to options (e. g. , camber or more reinforcement) that do not affect other members of design team.

Columns In braced frames, the moment in most columns will be less than the moment at minimum eccentricity. Exceptions will generally be in columns at or near the top story where minimum reinforcement can easily be increased during final design. Thus the required area of a column can be calculated by dividing the axial load by an average stress that includes the capacity of vertical reinforcement as well as the concrete.

For maximum economy, a low percentage of reinforcement should be used but not less than 0.5% to 1.0%. When reinforcement exceeds 4%, clearances should be checked to verify that column vertical bars can be placed and spliced with bars in columns above and below and that column bars passing through the floor framing will not interfere with bars in the floor.

The same size column can be maintained through 20 stories or more by increasing the concrete strength and using a high percentage of reinforcement in the lower stories. Fig. 12.1 illustrates how columns might be reinforced in a 20-story building. Only the top few stories would have columns with minimum reinforcement and excess capacity. If these columns were reduced in size, the extra cost of formwork would probably exceed the savings in concrete and steel. If a column size must be reduced, only one dimension should be changed (e.g., change from 24×24 in. to 24×18 in.).

Words and Expressions

iteration [ˌitəˈreiʃn]　*n.* 重复,迭代
ingredient [inˈgriːdjənt]　*n.* 配料
thaw [θɔː]　*v.* 融化
air entrainment　*n.* 加气剂
offset [ˈɔfset]　*n.* 偏移
flange [flændʒ]　*n.* 翼缘
taper [ˈteipə]　*v.* 逐渐变尖细

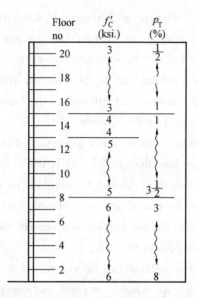

Fig. 12.1 Example of Concrete Strength and Reinforcement Ratio in Tall Building Columns

Passage B Project Cost Control

Introduction

Project is corporate image window and benefit of the source. With increasingly fierce market competition, the quality of work and the construction of civilizations rising material prices fluctuations. Uncertainties and other factors, make the project operational in a relatively tough environment. So the cost of control is through the building of the project since the bidding phase of acceptance until the completion of the entire process, It is a comprehensive enterprise cost management an important part, we must organize and control measures in height to the attention with a view to improving the economic efficiency of enterprises to achieve the purpose.

Outlining the Construction Project Cost Control

The cost of the project refers to the cost and process of formation occurred, on the production and operation of the amount of human resources, material resources and expenses, guidance, supervision, regulation and restrictions, in a timely manner to prevent, detect and correct errors in order to control costs in all project costs within the intended target, to guarantee the production and operation of enterprises benefits.

Construction Enterprise Cost Control Principle

Construction enterprises control the cost of control is based on cost control of construction project for the center, construction of the project cost control principle is the enterprise cost management infrastructure and the core, construction Project Manager in the Ministry of Construction of the project cost control process, we must adhere to the following basic principles.

(a) Principles lowest cost. Construction of the project cost control, the basic purpose is to cost management through various means, promote construction projects continue to reduce costs, to achieve the lowest possible cost of the objective requirements. The implementation of the principle of minimum cost, attention should be given to the possibility of reducing costs and reasonable cost of the minimum. While various mining capacity to reduce costs so that possibility into reality; the other must proceed from actual conditions, enacted subjective efforts could achieve a reasonable level of the minimum cost.

(b) Overall cost control principles. Cost Management is a comprehensive enterprise—wide and full management of the entire process, also known as the "three" of management. The full project cost control is a system of substantive content, including the departments, the responsibility for the network and team economic accounting, and so on; to prevent the cost control is everybody's responsibility, regardless of everyone. Project cost of the entire process control requirements to control its costs with the progress of construction projects in various stages of continuous, neither overlooked nor time when, should enable construction projects throughout costs under effective control.

(c) Dynamic Control principle. Construction of the project is a one-time, cost control should emphasize control of the project in the middle, that is, dynamic control. Construction preparation stage because the cost is under the control of construction design to determine the specific content of the cost, prepare cost plans, the development of a cost-control program for the future cost control ready. And the completion of phase cost control, as a result of cost financing has been basically a foregone conclusion, even if the deviation has been too late to rectify.

(d) Principle of management by objectives. Management objectives include setting goals and decomposition, the goal of responsibility and implementation of the aims of the inspection results of the implementation, evaluation of the goals and objectives that form the management objectives of the planning, implementation, inspection, processing cycle, PDCA.

(e) Responsibility, authority, in light of the profit principle. Construction of the project, project manager of the department, the team shouldering the responsibility for cost control at the same time, enjoy the power of cost control, project manager for the department, Teams cost control in the performance of regular examination and appraisal of implementation of a crossword punishment. Only to do a good job duty, rights, and interests combining cost control, in order to achieve the desired results.

The Construction Cost Control Measures

Cost control measures. Reduce the cost of construction projects means, we should not only increase revenue is also reducing expenditure, or both also increase savings. Cutting expenditure is not only revenue, or revenue not only to cut expenditure, it is impossible to achieve the aim of reducing costs, at least there is no ideal lower cost effective.

Project Manager of the project cost management responsibility for the first, comprehensive organization of the project cost management, timely understand and analyze profit and loss situation and take prompt and effective measures; engineering technology department should ensure the quality, Regular tasks to complete as much as possible under the premise adopt advanced technology in order to reduce costs; Ministry of Economic Affairs should strengthen budget management contract, the project to create the budget revenue; Finance Ministry in charge of the project's financial, Analysis of the project should keep the financial accounts of reasonable scheduling of funds. Develop advanced economies reasonable construction program, which can shorten the period, and improve quality, reduce costs purpose; paid attention to quality control to eliminate redone, shorten the acceptance and reduce expenses; control labor costs, material costs, Machinery and other indirect costs.

With the construction market competition intensifies, more and the price low, the scene increasingly high management fees. This requires project managers to more scientific and more rigorous management approach to the management of the project. As a management departments should be a reasonable analysis of regional economic disparities, to prevent the input across the board. From the foregoing analysis, project management and cost control are complementary, it is only by strengthening project management, can control project costs; only achieve cost control project aims to strengthen the management of construction project can be meaningful. Construction of the project cost control of construction reflects the nature of project management features, and represents construction project management at the core. Construction of the project cost control of construction project management performance evaluation of the objectivity and fairness of the scale.

Strengthen Project Cost Control Practical Significance

(a) Strengthen project cost control railway construction enterprises out of their predicament, the need to increase revenue. At present, the railway construction enterprises just into the market, to participate in market competition, will face a tough test of the market. Now the construction market liberalization, implement bidding system, and the strike has very low weight, to create efficiency is the only way to strengthen internal management and improve their internal conditions, internal efficiency potentials. Therefore, the strengthening of project cost control is a very realistic way.

(b) Strengthening Project Cost control is adapted to the market competition, and strengthening internal management to the needs of their work. With the railway enterprise's rapid development, construction increasingly fierce market competition. For a period of time, the railway construction enterprises will face the increasingly fierce market challenges Construction of the business environment difficult to be improved. Efficiency increases, effective cost control and claims will be strengthened in the future management focus. This requires the railway construction enterprises should respect the unity of the work to reduce costs and enhance efficiency objectives. In accordance with the requirements of the market economy research, adjustment and improve the management system, to further strengthen the management of infrastructure, enterprise management from the physical management to value management, thus enabling cost management into enterprise management centers.

Construction Project Cost Control Method

Construction cost control method of many, this highlights deviation analysis. Deviation refers to the actual value of the construction costs with the planned value of the difference. Deviation analysis may be used to a bar chart method, form method, curve method

(a) Bar Graph method is different Transverse-Line marking the completion of the project has been the construction costs, End to construction projects have been completed and cost (the cost-effective construction, Transverse-Line length is proportional to the amount of their cases. Bar Graph with image, audio-visual, very clear advantages, it can accurately express construction cost deviations, but one can feel the gravity of deviation. But this method is less amount of information

(b) Form method is error-analysis of the most commonly used method, it will project code name, construction of the cost parameters and construction cost deviation integrated into the number one form, and in the form of direct comparison. As the deviations are shown in the table, construction costs make integrated managers to understand and deal with these data. Flexible, applicability; informative; forms can be handled by computer, thus saving a large amount of data to deal with the human, and greatly improve speed.

(c) Curve is a total construction cost curve (S line curve) for the partial construction costs differential analysis methods. A figure which indicated the actual value of the construction cost curve, p. construction cost of the scheme said the value curve, the curve between two vertical distances between construction cost deviations. The method used is the same image analysis, and visual characteristics, but this is very difficult to direct for quantitative analysis of quantitative analysis can play a role.

Currently Construction Enterprise Project Cost Control Analysis

Problems and the causes of the current project implementation after the restructuring projects implemented "five responsibility for the costs", "100 of responsibility for the content of the output value of wages" and "contracting indicators" various forms of economic management contract responsibility system. Construction projects in the Ministry of Production and quality aspects of the rapid progress. But beyond doubt is just working, regardless of the mode of production accounts still exist. Some only production tasks are completed, the cost of a weak awareness cost management as dispensable. In the past two years the Department of grasping items complain, enterprise project appraisal of the indicators, all focus on the production tasks to complete, objective, fueled by such acts. Specific indications

(a) In the use of labor, not by post, according to the actual needs staffing, they can complete the work for three; can be used for low-cost trades and the use of subjects of labor costs. To take care of relations, sensibilities and TWA also retained his spare time. Workers can be indifferent to the production and operation, but the monthly wages, allowances, and bonuses can spend less. Artificially expand the expenditure of funds.

(b) Material management can be simplified to what extent on what is the level of simplification, operational staff only to facilitate easy and timely withdrawal credit card, and some kind of engineering materials and book a difference to thousands of dollars, tens of thousands or even hundreds of some of its few. Consuming the works, the procedure is incomplete. Not fixed by the material. Placing arbitrary site materials, engineering materials stolen have occurred from time to time; Consuming accessories not review, bad on the other, very few people to repair; fill empty fuel consumption result was secretly putting the oil sold.

(c) Construction machinery efficiency is not high (example Monthly leasing machinery and equipment), less to him, usually poor maintenance. With mechanical equipment failure analysis is not objective and subjective reasons, not to pursue the responsibility of the parties, have bad information on the exchange, no other information on. TWA did not undergo a rigorous examination on the induction training mechanical damage to the non-normal, impact of the construction progress.

In summary, the current project of cost management, accounting only after the accounting, rather than advance the prevention and control things. The reasons are lack of cost awareness. Simply that the cost of management is the financial sector or the superior leadership, have nothing to do with them. Only focused on the "production tasks are completed" and "contracting profit and loss", the groups have a "negative effect". Therefore, project to mobilize the full participation of the Ministry of cost control, deepening of the project cost management imperative.

The construction project cost control is a complicated systematic project. The application needed to be applied with flexibility the actual operation is adapted to local conditions, different sizes, different construction firms and different management systems have differences, but in any case are the construction of the production and operation of enterprises in the amount of human resources, material resources and expenses, guidance, supervision, regulation and restriction. Therefore, "increases production and economize, to increase revenues and reduce expenditures" is a common construction enterprises, this requires constant practice in the review and improve cost control, ways and means to ensure that the project cost goals. As an enterprise only deepening financial management system, outstanding cost management center, further strengthening cost management and strict cost veto, the full implementation, the whole process comprehensive cost control and continually adapt to the market competition situation, overcome adversity, to achieve the goal of cost control.

Words and Expressions

revenue ['revənjuː] n. 税收,收益,收入
flexibility [ˌfleksə'biliti] n. 柔韧性,机动性,灵活性
project manager 项目经理
project cost control 项目成本控制/管理
cost management 造价管理,成本管理

Dialogue

A: I'd like to know the measures you've taken to ensure the quality control of the reinforcement works.
B: First we had a visit to the steel manufacturer where we inspected their production procedures and quality control method. Satisfied with the quality of the products, we placed the material order with the approval of the Employer.
A: What did you do after the reinforcement bars arrived on site?
B: We checked the size and surface and we had rebar tests in our laboratory.
A: What kind of tests did you do?

B: We had tests for tension strength, extension rate, and cold bending degree.

A: What's the result?

B: Our tests show that the reinforcement bars are of good quality. These are the Quality Certificates and Test Reports from the manufacturer. And this is our Test Report. All physical qualities and chemical components are in compliance with the specification.

Words and Expressions

reinforcement works　钢筋工程

inspect [in'spekt] v. 监督,监理,监察

place an order (for ... with)　(向……)订货

rebar test　钢筋试验

tension strength　抗拉强度

extension rate　延伸率

cold bending degree　冷弯曲度

Quality Certificates　质量合格证书

physical qualities　物理性能

chemical components　化学成分

Unit 13

Passage A Optimization

General

The optimum structural frame is the best, or most favorable, in terms of an objective, from among possible satisfactory frames. "Optimization" is the process of finding the optimum frame. The term optimization is frequently used to mean the process of finding the least cost frame, but engineers must also optimize other objectives to achieve success in design practice.

Because a structural frame cannot be the optimum frame for all objectives, some balancing of conflicting objectives is required. For example, a frame cannot have long spans and also least cost. Likewise, minimum floor thickness is not consistent with small deflection. To help resolve conflicting objectives, consider the following procedure:

1. Assign priorities to objectives. Put yourself in the position of the architect, the owner, and other people in the design and construction team and visualize how they would judge each objective,

even if their judgment is subjective rather than objective.

2. Define and clarify the issues. Which objective can be optimized and which cannot? In the range of practical solutions, which objectives will be affected most?

3. Develop options that emphasize one or more objectives. Such options are the only ones likely to be accepted.

4. Evaluate how well each option or possible structural frame meets each objective and prepare your recommendation for the frame to be used.

Before discussing optimization further, it is advisable to consider structural optimization in the context of the overall optimization hierarchy. Society and the building industry will optimize a project at several stages before completion.

The society of a community, state, or nation will influence whether to build a facility, how big it should be, for what purpose it should be built, and other such broad questions. Society makes these decisions through laws and peer pressure. In addition to government and a community's society, the owner of a proposed facility will optimize a facility in relation to these same questions.

The location of a facility will be optimized by the owner to minimize operating costs and maximize operating benefits. Retail facilities will be located near customers, schools will be located conveniently for children, and so forth. Sometimes realtors and architects assist owners in the decision on where to locate a facility. Sometimes geotechnical and structural engineers will assist in this decision when soil conditions at potential sites materially affect construction cost or constructibility, especially if the effect is adverse.

Building layout and general arrangement will be optimized, generally by the architect and the owner. The structural engineer may influence the design of the building at this stage, especially if the structural frame is an important feature of the building (as in a long-span roof or structural frame used for its aesthetic appeal) or if the cost of the frame is a large part of the total cost (as in a warehouse, a parking garage, or a tall building).

The structural frame will be optimized by the structural engineer during preliminary design with some assistance from other people in the design and construction team.

After a structural frame has been selected, the final design should optimize, which generally means minimize, the construction cost. Because formwork costs and concrete quantities are established in preliminary design, minimizing construction costs usually means reducing the amount of reinforcing steel used. To a lesser extent, other objectives may also be optimized during detailed design. Note that structural safety is not optimized as it is an absolute requirement. Only occasionally will an owner request a building with greater-than-normal factors of safety.

The above hierarchy has been arranged in decreasing order of impact on cost and probably on other objectives as well. Social concerns, location, and layout of a building can each affect its life cycle coats by 100% or more. Preliminary design can easily affect the cost of a structural frame by 30% to 50% whereas minimizing the amount of reinforcing steel used in the final design can affect the construction costs by only a few percent. Obviously then, engineers should concentrate their optimizing and cost-saving efforts on the preliminary design.

(1) Construction of the frame.

(2) Construction of the remainder of the building.

(3) Financing, legal, and administrative costs during and after construction.

(4) Maintenance and repair.

(5) Remodeling.
(6) Demolition.
(7) Salvage.

Cost of Optimization

Even neglecting other objectives, optimum or minimum cost rarely coincides with minimum material quantities, but most technical papers on the subject make this assumption. Indeed, it is extremely difficult to reduce optimization to a mathematical process because the total cost of the frame includes the entire structure and not just the member or group of members under consideration. The structural frame is only part of a project. The cost of other parts of the project are affected by the structural frame and by individual structural members.

Minimizing the cost of construction is an important criterion in selecting a framing scheme. Engineers can usually estimate the cost of construction with satisfactory qualitative, if not quantitative, accuracy, by consulting local constructors and material suppliers on unit costs of material and labor of erection. Care must be taken to consider factors affecting unit costs by soliciting such information from contractors and suppliers. To make impartial cost comparisons, the engineer must be consistent between schemes in estimating quantities and applying unit costs.

Include all costs in each scheme even through they may be identical from one scheme to another, such as floor finishing, and include a reasonable allowance for the contractor's overhead and profit. These will lend credibility to the engineer's cost estimate and increase the chances of its acceptance. The total cost will be easier to compare to cost estimates prepared by the architect, contractor, or other people.

Make allowance in the cost estimate for nontypical panels. For example, the average number of columns in one panel will vary from four to a little more than one, depending upon the number of bays in the width and length of the building. Special conditions around the perimeter or at the stair and elevator shafts can usually be disregarded because they will affect all schemes in a similar manner.

If the thickness of the floor framing varies between schemes, allowance should be made for the cost of walls, stairs, elevators, mechanical risers, and other items of construction that vary with floor-to-floor heights.

To understand how to minimize construction costs, it is essential to know the distribution of costs for the various elements of construction and to what degree an engineer might influence these costs. This understanding comes only through the preparation of costs analyses on many structural frames. Table 1 has been prepared to illustrate a range of costs for typical concrete frames based on average costs in 1985. Some projects will fall below or above the ranges for both material quantities and costs. The cost of formwork is usually 50% to 60% of the total cost of the structural frame but could be as little as 35% in exceptional cases. For a particular project, engineers should prepare more detailed cost estimates than those shown in Tab. 13.1 to guide their optimization efforts.

The cost of reinforcing steel is minimized by using grade 60 bars, by increasing the size and reducing the number of bars to be placed, by minimizing the amount of bending required, by using standard details, and by repetition of bars exactly alike in size, length, and bending. Avoid #14 and #18 bars as they are not usually carried in stock but are rolled on special order. Use lap splices rather than welded or mechanical splices. Place bars in a single layer in beams. Use the largest practical stirrup at standard spacing. Check the fit and clearances of all bars. A well-conceived preliminary design will allow such optimization to take place. Welded wire fabric (WWF) or prefabricated mats might reduce placing costs.

Tab. 13.1 Construction Cost Allocation and Opportunities for Cost Reduction

Item	Quantities (per ft)	Cost ($ per ft)	Influence of Structural Engineer	
			percent	Amount ($ per ft^2)
Finishing	1.0 ft^2	0.3 to 0.4	0	0
Steel	2.5 to 12 lb	0.4 to 2.3	20	0.5
Concrete	0.5 to 1.0 ft^3	0.9 to 3.00	20	0.6
Formwork	1.05 to 2.0 ft^2	1.4 to 5.00	50	2.50
Total		3.00 to 9.20		3.6

This table uses approximate, average 1985 prices and approximate quantities for illustration purposes only. It is not intended as an accurate statement of probable construction costs.

The cost of steel is also minimized by reducing the quantity. This is best accomplished by selecting bar combinations as close to the required quantity as possible. Reducing the quantity of steel by cutting off bars within the span of flexural members requires much calculation effort and risks construction errors that could have serious consequences. Furthermore, the potential saving in steel is small.

Avoid spiral columns where possible, as the weight of steel is greater than that in tied columns except for large round columns with low concrete strength and few vertical bars. Also, fabrication of spirals costs more than fabrication of vertical bars.

Bundled bars in columns with more than four vertical bars may make placing concrete easier, reduce interference with intersecting beam bars, and reduce construction cost.

The cost of concrete is minimized by reducing the number of mix designs required, specifying aggregate size and mix designs appropriate for structural members with congested reinforcement, and limiting strength and other requirements to values that can be reasonably accomplished by local concrete suppliers. It is sometimes necessary and desirable to challenge concrete suppliers to furnish concrete of higher strength than they have furnished before, but the cost will also likely be higher. The quantity of concrete should be minimized on preliminary design. In final design, there is little or no flexibility in reducing concrete quantities.

The cost of concrete placement will be minimized if the engineer considers problems in placing and consolidating concrete. Because it is difficult to cast concrete through a curtain of closely spaced top reinforcement, periodic gaps of 6 to 8 in. should be provided to allow insertion of pumping hoses and vibrators. Making it easier to place concrete helps assure its high quality as well as reduce costs. Such gaps in reinforcement should be provided, even though the bars between gaps must be spaced closer, or bundled, or placed in a second layer.

Because smaller quantities are involved, because placing is more difficult and higher strength concrete is more likely to be used, the cost of concrete in columns is more than the cost of concrete in floor slabs. The cost of column concrete can be reduced by using fewer but larger columns and by specifying ties that permit an unobstructed area in the center of the column for placing concrete. Although high-strength concrete costs more than concrete of moderate strength, the increase in cost is less than the increase in strength. Thus, high-strength concrete in columns is economical if it can be fully stressed.

The cost of formwork is minimized by simplifying and repeating the shapes to be formed as much as possible and by reducing the surface area of formwork. the suggestions below will help achieve these goals.

(1) Space columns uniformly. Uniform loading on slabs, beams, and columns is likely to result, and sizes of these members can be made uniform. Repetition will lower construction costs even though a little extra material may be required.

(2) Use one column size from foundation to roof and for as many columns in each story as possible. By doing so, form building is simplified, column forms can be reused, column ties will be standardized, deck forms around the column will not need to be reworked, fewer carpenter and ironworker foremen will be required, less labor will be used, fewer supervisors and inspectors will be needed, and costs will be lower.

(3) Use one wall thickness from foundation to top.

(4) Use steel- or fiber- reinforced plastic forms where the improved surface texture will permit omission of other, more expensive finish materials.

(5) Use shear walls or other bracing systems to reduce the size of columns.

(6) Make many beams the same size even though the spans and loading different. Use more reinforcing steel or less, as required to carry the load. Beams of uniform size are less expensive for many of the same reasons listed above for columns. Beam sizes need change only if loads and moments change by 30% to 50% or more.

(7) Make beams at least 2 in. wider than columns on each side, as shown in Fig. 14.4, so that column and beam bars pass each other without interference. Beams should not be narrower than columns because forms for the column-beam connection will be expensive.

(8) Use wide shallow beams to minimize floor-framing depth. This will simplify installation of mechanical piping, duct work, and equipment without interference with the structural frame. Coordination

between trades takes everyone's time—architects, engineers, contractors, and suppliers. Lower floor-to-floor heights will reduce the cost of walls and vertical building elements, such as stairs and elevators. It is easier to place steel in a wide, shallow beam than in a deep, narrow beam. Furthermore, pipes and even duct work can pass vertically through a wide beam, whereas such penetrations may be impossible in a narrow beam. Rerouting pipes and ducts around beams is part of the cost of construction.

(9) Use one deck depth, as in a flat plate or in a pan slab, where beams are the same depth as joists. The contractor can use one length of shoring, and fewer problems will be experienced in coordinating concrete construction with mechanical trades.

(10) Consider "flying forms" if the structural frame is large and many typical bays can be made exactly the same. A flying form is a form for an entire bay or portion of a bay built as a unit, lowered from the hardened concrete slab after the slab has gained sufficient strength, and moved or flown to the next higher level. At least 8 or 10 reuses of each form are necessary so that the savings of labor in reuse will exceed the extra cost of the form. For maximum efficiency, the entire building slab should be formed with flying forms.

(11) Consider stock lumber and plywood sizes in sizing beams and columns. Stock materials allow immediate starts, reduce worker learning curves, reduce jobsite errors and lower material costs.

(12) For pan slabs and waffle slabs, use standard stock sizes of pans and domes.

(13) Use one size pan throughout a project. (Pricing of pan slab forms is based on the number of reuses and the floor area, not on the surface contact area of the pans)

(14) Use no coffers or recesses in the underside of a slab unless there is a substantial advantage in doing so. The amount of concrete

saved usually will not offset the additional cost of formwork.

(15) If depressions are required in the top of the slab for terazzo, ceramic tile, or other floor finishes, keep the bottom of the slab flush, at least in the bay in which depressions are required.

(16) Eliminate drop panels in a flat slab and use a flat plate instead. If column capitals are required, make all capitals the same size.

(17) Avoid offsets, brackets, corbels, pilasters, haunches, and all other interruptions to the smooth surface of formwork.

(18) Avoid small changes in concrete outline. Instead, make up differences in finish materials, such as masonry, partitions, and floor toppings.

(19) Allow for reasonable tolerances in constructing concrete so that windows, doors, walls, partitions, and other abutting construction will fit with ease.

(20) Plan the procedures for reshoring to minimize the amount required and simplify its placement.

Making a project simpler to build not only reduces costs but also helps ensure that fewer construction mistakes will be made. Fewer mistakes mean that the project is more likely to be built according to the engineer's intent.

Words and Expressions

subjective [sʌb'dʒektiv]　*a.* 客观的
objective [əb'dʒektiv]　*a.* 主观的
clarify ['klærifai]　*v.* (弄)明白
hierarchy ['haiərɑːki]　*n.* 体系
retail ['riːteil]　*n.* 零售
adverse ['ædvəːs]　*a.* 有害的
aesthetic [iːs'θetik]　*a.* 美学的

solicit [sə'lisit] v. 请求协助
flush [flʌʃ] a. 平的,平齐的
offset ['ɔfset] n. 偏移
corbel ['kɔːbəl] n. 梁托,牛腿
pilaster [pi'læstə] n. 壁柱
haunch [hɔːntʃ] n. 梁腋
tolerance ['tɔlərəns] n. 公差
abut [ə'bʌt] v. 连接

Passage B Risk Analysis of the International Construction Project

Abstract

This analysis used a case study methodology to analysis the issues surrounding the partial collapse of the roof of a building housing the headquarters of the Standards Association of Zimbabwe (SAZ). In particular, it examined the prior roles played by the team of construction professionals. The analysis revealed that the SAZ's traditional construction project was generally characterized by high risk. There was a clear indication of the failure of a contractor and architects in preventing and/or mitigating potential construction problems as alleged by the plaintiff. It was reasonable to conclude that between them the defects should have been detected earlier and rectified in good time before the partial roof failure. It appeared justified for the plaintiff to have brought a negligence claim against both the contractor and the architects. The risk analysis facilitated, through its multi-dimensional approach to a critical examination of a construction problem, the identification of an effective risk management strategy for future construction projects. It further served to emphasize the point that clients are becoming more demanding, more discerning, and less willing to accept risk without recompense. Clients do not want surprise, and are more likely to engage in litigation when things go wrong.

The structural design of the reinforced concrete elements was done by consulting engineers Knight Piesold (KP). Quantity surveying services were provided by Hawkins, Leshnick & Bath (HLB). The contract was awarded to Central African Building Corporation (CABCO) who was also responsible for the provision of a specialist roof structure using patented "gang nail" roof trusses. The building construction proceeded to completion and was handed over to the owners on Sept. 12, 1991. The SAZ took effective occupation of the headquarters building without a certificate of occupation. Also, the defects liability period was only three months.

The roof structure was in place 10 years before partial failure in December 1999. The building insurance coverage did not cover enough, the City of Harare, a government municipality, issued the certificate of occupation 10 years after occupation, and after partial collapse of the roof.

At first the SAZ decided to go to arbitration, but this failed to yield an immediate solution. The SAZ then decided to proceed to litigate in court and to bring a negligence claim against CABCO. The preparation for arbitration was reused for litigation. The SAZ's quantified losses stood at approximately $ 6 million in Zimbabwe dollars (US $1.2 m).

After all parties had examined the facts and evidence before them, it became clear that there was a great probability that the courts might rule that both the architects and the contractor were liable. It was at this stage that the defendants' lawyers requested that the matter be settled out of court. The plaintiff agreed to this suggestion, with the terms of the settlement kept confidential.

The aim of this critical analysis was to analyse the issues surrounding the partial collapse of the roof of the building housing the HQ of Standard Association of Zimbabwe. It examined the prior roles played by the project management function and construction professionals in preventing/mitigating potential construction problems. It further assessed the extent to which the employer/client and parties to a construction contract are able to recover damages under that contract. The main objective of this critical analysis was to identify an effective risk management strategy for future construction projects. The importance of this study is its multidimensional examination approach.

Experience suggests that participants in a project are well able to identify risks based on their own experience. The adoption of a risk management approach, based solely in past experience and dependant on judgement, may work reasonably well in a stable low risk environment. It is unlikely to be effective where there is a change. This is because change requires the extrapolation of past experience, which could be misleading. All construction projects are prototypes to some extent and imply change. Change in the construction industry itself suggests that past experience is unlikely to be sufficient on its own. A structured approach is required. Such a structure can not and must not replace the experience and expertise of the participant. Rather, it brings additional benefits that assist to clarify objectives, identify the nature of the uncertainties, introduces effective communication systems, improves decision-making, introduces effective risk control measures, protects the project objectives and provides knowledge of the risk history.

Construction professionals need to know how to balance the contingencies of risk with their specific contractual, financial, operational and organizational requirements. Many construction professionals look at risks in dividually with a myopic lens and do not realize the potential impact that other associated risks may have on their business operations. Using a holistic risk management approach will enable a firm to identify all of the organization's business risks. This will increase the probability of risk mitigation, with the ultimate goal of total risk elimination .

Recommended key construction and risk management strategies for future construction projects have been considered and their explanation follows. J. W. Hinchey stated that there is and can be no 'best practice' standard for risk allocation on a high-profile project or for that matter, any project. He said, instead, successful risk management is a mind-set and a process. According to Hinchey, the ideal mind-set is for the parties and their representatives to, first, be intentional about identifying project risks and then to proceed to develop a systematic and comprehensive process for avoiding, mitigating, managing and finally allocating, by contract, those risks in optimum ways for the particular project. This process is said to necessarily begin as a science and ends as an art.

According to D. Atkinson, whether contractor, consultant or promoter, the right team needs to be assembled with the relevant multi-disciplinary experience of that particular type of project and its location. This is said to be necessary not only to allow alternative responses to be explored. But also to ensure that the right questions are asked and the major risks identified. Heads of sources of risk are said to be a convenient way of providing a structure for identifying risks to completion of a participant's part of the project. Effective risk management is said to require a multi-disciplinary approach. Inevitably risk management requires examination of engineering, legal and insurance related solutions.

It is stated that the use of analytical techniques based on a statistical approach could be of enormous use in decision making.

Many of these techniques are said to be relevant to estimation of the consequences of risk events, and not how allocation of risk is to be achieved. In addition, at the present stage of the development of risk management, Atkinson states that it must be recognized that major decisions will be made that can not be based solely on mathematical analysis. The complexity of construction projects means that the project definition in terms of both physical form and organizational structure will be based on consideration of only a relatively small number of risks. This is said to then allow a general structured approach that can be applied to any construction project to increase the awareness of participants.

The new, simplified Construction Design and Management Regulations (CDM Regulations) which came in to force in the UK in April 2007, revised and brought together the existing CDM 1994 and the Construction Health Safety and Welfare (CHSW) Regulations 1996, into a single regulatory package.

The new CDM regulations offer an opportunity for a step change in health and safety performance and are used to reemphasize the health, safety and broader business benefits of a well-managed and co-ordinated approach to the management of health and safety in construction. I believe that the development of these skills is imperative to provide the client with the most effective services available, delivering the best value project possible.

Construction Management at Risk (CM at Risk), similar to established private sector methods of construction contracting, is gaining popularity in the public sector. It is a process that allows a client to select a construction manager (CM) based on qualifications; make the CM a member of a collaborative project team; centralize responsibility for construction under a single contract; obtain a bonded guaranteed maximum price; produce a more manageable, predictable project; save time and money; and reduce risk for the client, the architect and the CM.

CM at Risk, a more professional approach to construction, is taking its place along with design-build, bridging and the more traditional process of design-bid-build as an established method of project delivery.

The AE can review the CM's approach to the work, making helpful recommendations. The CM is allowed to take bids or proposals from subcontractors during completion of contract documents, prior to the guaranteed maximum price (GMP), which reduces the CM's risk and provides useful input to design. The procedure is more methodical, manageable, predictable and less risky for all.

The procurement of construction is also more business-like. Each trade contractor has a fair shot at being the low bidder without fear of bid shopping. Each must deliver the best to get the project. Competition in the community is more equitable all subcontractors have a fair shot at the work.

A contingency within the GMP covers unexpected but justifiable costs, and a contingency above the GMP allows for client changes. As long as the subcontractors are within the GMP they are reimbursed to the CM, so the CM represents the client in negotiating inevitable changes with subcontractors.

There can be similar problems where each party in a project is separately insured. For this reason a move towards project insurance is recommended. The traditional approach reinforces adversarial attitudes, and even provides incentives for people to overlook or conceal risks in an attempt to avoid or transfer responsibility.

A contingency within the GMP covers unexpected but justifiable costs, and a contingency above the GMP allows for client changes. As long as the subcontractors are within the GMP they are reimbursed to the CM, so the CM represents the client in negotiating inevitable changes with subcontractors.

There can be similar problems where each party in a project is separately insured. For this reason a move towards project insurance is recommended. The traditional approach reinforces adversarial attitudes, and even provides incentives for people to overlook or conceal risks in an attempt to avoid or transfer responsibility.

It was reasonable to assume that between them the defects should have been detected earlier and rectified in good time before the partial roof failure. It did appear justified for the plaintiff to have brought a negligence claim against both the contractor and the architects.

In many projects clients do not understand the importance of their role in facilitating cooperation and coordination; the design is prepared without discussion between designers, manufacturers, suppliers and contractors. This means that the designer can not take advantage of suppliers' or contractors' knowledge of build ability or maintenance requirements and the impact these have on sustainability, the total cost of ownership or health and safety.

This risk analysis was able to facilitate, through its multi-dimensional approach to a critical examination of a construction problem, the identification of an effective risk management strategy for future construction projects. This work also served to emphasize the point that clients are becoming more demanding, more discerning, and less willing to accept risk without recompense. They do not want surprises, and are more likely to engage in litigation when things go wrong.

Words and Expressions

risk [risk]　　*n.* 风险,危险,冒险

dividually [di'vidjuəli]　　*adv.* 分开地,分享地,可分割地

co-ordinate [kəu'ɔːdənit]   v. 使……协调

Dialogue

A: OK. We will check the equipment cost and see if there are any costs that can be deducted. The Item 4 cost is the material cost which is the largest one and occupy 55% of the total cost.

B: From the breakdown we see the cement cost USD 4.97 million. How much is the purchasing price for each tone of cement.

A: From our investigation the cement is USD 70 per tone including the cost of material, transportation and other costs. We have already underestimated this cost. The risk of inflation in next three years has not been considered yet.

B: It seems that 71 000 ton of cement for around 220 000 m^3 concrete is higher than usual.

A: We can not judge this item by average calculation. There are several grades of concrete for different parts of the Works and the mixture of each grade of concrete is different. And also we should think of the consumable loss during delivery and operation.

B: We had better have a detailed breakdown for concrete, reinforcement and formwork in latter stage.

Words and Expressions

underestimated [ˌʌndər'estimeitid]   v. 低估
equipment cost   设备费
material cost   材料费
risk of inflation   通货膨胀风险

Appendix 1

建筑工程常用英语词汇

1. Name of Professional Role 职务名称

监理单位 consultant company
总经理 general manager
项目经理 project manager
项目副经理 project deputy manager
商务经理 business/commercial manager
采购经理 purchasing manager
质量工程师 quality engineer
结构工程师 structure engineer
首席建筑师/设计师 principal architect
建筑师/设计师 architect
总工程师 chief engineer
土木工程师 civil engineer
工艺工程师 process engineer
电气工程师 electrical engineer
维修工程师 maintenance engineer
实验测试员 laboratory & testing engineer
机械设备工程师 machinery & equipment engineer
助理工程师 assistant engineer
会计主管 accountant supervisor
行政经理 administration manager
后勤主管 logistics supervisor

管理顾问 management consultant
预算员 quantity surveyor
测量员 surveyor
操作手 operator
工程技术员 engineering technician
安全员 safety engineer
翻译 translator
保安 security guider
巡逻人员 patrol
实习生 trainee
制图员 draftsman

2. Discipline 专业

建筑 architecture
土木 civil
给排水 water supply and drainage
总图 plot plan
采暖通风 H. V. A. C (heating、ventilation and air conditioning)
电力供应 electric power supply
电气照明 electric lighting
电讯 telecommunication
仪表 instrument
热力供应 heat power supply
动力 mechanical power
工艺 process technology
管道 piping

3. Drafting 制图

总说明 general specification
工程说明 project specification

采用标准规范目录 list of standards and specification adopted
图纸目录 list of drawings
图纸目录及说明 list of contents and description
建筑施工图 architectural working drawing
建筑图 architectural drawing
结构图 structural drawing
总图 general plan
平面图 plan
基础平面图 foundation plan
底层平面图 ground floor plan
标准层平面图 standard floor plan
顶层平面图 top floor plan
立面图 elevation drawing
正面图 front elevation
背立面图 back elevation
侧面图 side elevation
右立面图 right elevation
详图,大样图 detail drawing
留孔平面图 plan of provision of holes
截(剖、断)面图 sectional drawing
剖面 section
纵剖面 longitudinal section
横剖面 cross (transverse) section
立面 elevation
正立面 front elevation
透视图 perspective drawing
侧立面 side elevation
背立面 back elevation
示意图 diagram
草图 sketch

荷载简图 load diagram
流程示意图 flow diagram
标准图 standard drawing
地形图 topographical map
土方工程图 earth-work drawing
模板图 formwork drawing
首页 front page
图例 legend
净空 clearance
净高 headroom
净距 clear distance
净跨 clear span
截面尺寸 sectional dimension
开间 bay
进深 depth
单跨 single span
双跨 double span
多跨 multi-span
标高 elevation, level
绝对标高 absolute elevation
设计标高 designed elevation
室外地面标高 ground elevation
室内地面标高 floor elevation
柱网 column grid

4. Project Construction Cost 工程造价

估算/费用估算 estimate/cost estimate
估算类型 types of estimate
详细估算 defined estimate
设备估算 equipment estimate

分析估算 analysis estimate
报价估算 proposal estimate
控制估算 control estimate
初期控制估算 interim control estimate/initial control estimate
批准的控制估算 initial approved cost
核定估算 check estimate
首次核定估算 first check estimate
二次核定估算 production check estimate
人工时估算 man hour estimate
材料费用/直接材料费用 material cost/direct material cost
设备费用/设备购买费用 equipment cost/purchased cost of equipment
散装材料费用/散装材料购买费用 bulk material cost/purchased cost of bulk material
施工费用 construction cost
施工人工费用 labor cost/construction force cost
设备安装人工费用 labor cost associated with equipment
散装材料施工安装人工费用 labor cost associated with bulk materials
人工时估算定额 standard man hours
施工人工时估算定额 standard labor man hours
标准工时定额 standard hours
劳动生产率 labor productivity/productivity factor/productivity ratio
施工监督费用 cost of construction supervision
施工间接费用 cost of construction indirect
分包合同费用/现场施工分包合同费用 subcontract cost/field subcontract cost
公司本部费用 home office cost
公司管理费用 overhead

非工资费用 non payroll
开车服务费用 cost of start-up services
其他费用 other cost
利润/预期利润 profit/expected profit
服务酬金 service gains
风险 risk
风险分析 risk analysis
风险备忘录 risk memorandum
未可预见费 contingency
基本未可预见费 average contingency
最大风险未可预见费 maximum risk contingency
用户变更/合同变更 client change/contract change
认可的用户变更 approved client change
待定的用户变更 pending client change
项目变更 project change
内部变更 internal change
批准的变更 authorized change
强制性变更 mandatory change
选择性变更 optional change
内部费用转换 internal transfer
认可的预计费用 anticipated approved cost
涨价值 escalation
项目费用汇总报告 project cost summary report
项目实施费用状态报告 project operation cost status report
总价合同 lump sum contract
偿付合同 reimbursable contract
预算 budget
财务 financial
竣工 completion
决算 final

工程 project
造价 construction cost
财务竣工决算审计 Financial completion final accounts audit
工程造价审计 project construction cost audit

5. Concrete Engineering 混凝土工程

砼配合比 concrete coordinating proportion
砼外加剂 addition mixed with concrete
早强剂 early-strength admixture
减水剂 water-reducing admixture ; drying agent
缓凝剂 retarder
搅拌时间 mixing duration
砼养护 concrete curing
砼运输 concrete transportation
砼运输车 concrete buggy
砼浇灌车 concrete-pouring vehicle
缺少砼 lack of concrete
水泥标号 cement mark
砼强度/标号 strength of concrete
砼抗压强度 compressive strength of concrete
砼抗拉强度 tensile strength of concrete
高强(度)砼 high-strength concrete
防水砼 waterproof concrete
耐酸砼 acid-resisting concrete
耐碱砼 alkaline-resisting concrete
沥青砼 asphalt concrete
泡沫砼 foam concrete
钢筋混凝土 reinforced concrete
轻质混凝土 lightweight concrete
细石混凝土 fine aggregate concrete

沥青混凝土 asphalt concrete
泡沫混凝土 foamed concrete
炉渣混凝土 cinder concrete
素混凝土 plain concrete/non-reinforced concrete
浇注 pouring
浇注混凝土 concreting

6. Steel Bar and Processing 钢筋及加工

配筋 arrangement of reinforcement
钢筋接头 steel bar joint
钢筋锚固 steel bar anchor
最小配筋率 minimum ratio of reinforcement
受力筋 carrying bar
分布钢筋 distributing bars/ distribution steel
主钢筋 main bar
辅助钢筋 auxiliary bars
构造配筋 construction reinforcement
纵向受力钢筋 longitudinal carrying bar
受拉钢筋 tensioned reinforcement
箍筋 hoop reinforcement
箍筋间距 stirrup spacing
抗剪钢筋 shear reinforcement
抗扭钢筋 turn-resisting steel bar
钢筋代换 steel bar replacement
钢筋除锈 get the rust of steel bar off
钢筋切断 steel bar cutting off
钢筋弯曲 bar bending
冷拉(拔)钢筋 cold drawn bar
钢筋搭接 bar splicing
钢筋焊接 steel bar welding

对接焊 butt-weld
搭接焊 lap welding
钢筋绑扎 steel bar binding
钢筋网 reinforcing fabric
预应力钢筋 prestressed reinforcing bar
预留钢筋 reserved steel bar
圈梁钢筋 girth reinforcement
钢筋保护层 cover to reinforcement
钢丝 steel wire
弯起钢筋 bent-up bar
双向配筋 two-way reinforcement
配筋率 reinforcement ratio
配箍率 stirrup ratio

7. Ground Base and Foundation 地基及基础

刚性基础 rigid foundation
柔性基础 flexible foundation
条形基础 strip foundation
独立基础 isolated foundation, individual foundation
筏式基础 raft foundation
箱形基础 box foundation
端承桩 end-bearing pile
摩擦桩 friction pile, floating pile
群桩 grouped piles
板桩 sheet pile, sheeting pile
板桩基础 sheet pile foundation
基础埋深 embedded depth of foundation
毛石基础 rubble foundation
阶形基础 stepped foundation
联合基础 combined foundation

饱和黏土 saturation clay
冰冻线 frost line, freezing level
不均匀沉降 unequal settlement, differential settlement
沉降 settlement
沉降差 difference in settlement
沉降缝 settlement joint
持力层 bearing stratum
挡土墙 retaining wall, breast wall
底板 base slab, base plate, bed plate
地板 floor board
地基 ground base, ground
地基承载力 ground bearing capacity
地基处理 ground treatment, soil treatment
地梁 ground beam, ground sill
地下工程 substructure work, understructure work
地下室 basement, cellar
地下水 ground water
地下水位 groundwater level, water table
地下水压力 ground water pressure
地质报告 geologic report
垫层 bedding, blinding
覆土 earth covering
固结 consolidation
灌注桩 cast-in-place pile, cast in site pile
护坡 slope protection, revetment
灰土 lime earth
回填 backfill, backfilling
回填土 backfill, backfill soil
基槽 foundation trench
基础 foundation, base

基础底板 foundation slab
地基勘探 site exploration, site investigation
基坑 foundation pit
集水坑 collecting sump
井点 well point
开挖 excavation, cutting
勘测 exploration and survey
勘测资料 exploration data
埋置 embedment
密实度 compactness, density, denseness
黏土 clay
黏质粉土 clay silt
碾压 roller compaction, rolling
排水 drainage, dewatering
排水沟 drainage ditch
排水孔 weep hole, drain hole
排水设备 dewatering equipment
容许沉降 permissible settlement
容许承载力 allowable bearing
软土 soft soil
砂垫层 sand bedding course, sand cushion
砂土 sandy soil, sands
砂质粉土 sandy silt
设备基础 equipment foundation
水泥搅拌桩 cement injection
素土夯实 rammed earth, packed soil
碎石桩 stone columns
弹性地基梁 beam on elastic foundation
土方工程 earthwork
挖方 excavation work, excavation

填方 fill, filling
压实 compaction, compacting
压实系数 compacting factor
验槽 check of foundation subsoil
预制混凝土桩 precast concrete pile
中砂 medium sand
桩承台 pile cap
钻孔桩 bored pile
钻探 exploration drilling, drilling,
最终沉降 final settlement

8. Wall 墙体

砖墙 brick wall
砌块墙 block wall
清水砖墙 brick wall without plastering
抹灰墙 rendered wall
石膏板墙 gypsum board, plaster board
空心砖墙 hollow brick wall
承重墙 bearing wall
非承重墙 non-bearing wall
纵墙 longitudinal wall
横墙 transverse wall
外墙 external (exterior) wall
内墙 internal (interior) wall
填充墙 filler wall
防火墙 fire wall
窗间墙 wall between window
空心墙 cavity wall
压顶 coping
圈梁 gird, girt, girth

玻璃隔断 glazed wall
防潮层 damp-proof course（D.P.C）
遮阳板 sunshade
阳台 balcony
伸缩缝 expansion joint
沉降缝 settlement joint
抗震缝 seismic joint
复合夹心板 sandwich board
压型单板 corrugated single steel plate
外墙板 cladding panel
复合板 composite panel
轻质隔断 light-weight partition
牛腿 bracket
砖烟囱 brick chimney
勒脚（基座）plinth

9. Floor and Trench 地面及地沟

地坪 grade
地面和楼面 ground and floor
素土夯实 rammed earth
炉渣夯实 tamped cinder
填土 filled earth
回填土夯实 tamped backfill
垫层 bedding course, blinding
面层 covering, finish
结合层 bonding (binding) course
找平层 leveling course
素水泥浆结合层 neat cement binding course
混凝土地面 concrete floor
水泥地面 cement floor

机器磨平混凝土地面 machine toweled concrete floor
水磨石地面 terrazzo flooring
马赛克地面 mosaic flooring
瓷砖地面 ceramic tile flooring
油地毡地面 linoleum flooring
预制水磨石地面 precast terrazzo flooring
硬木花地面 hard-wood parquet flooring
搁栅 joist
硬木毛地面 hard-wood rough flooring
企口板地面 tongued and grooved flooring
防酸地面 acid-resistant floor
钢筋混凝土楼板 reinforced concrete slab (R. C Slab)
乙烯基地面 vinyl flooring
水磨石嵌条 divider strip for terrazzo
地面做2%坡 floor with 2% slope
集水沟 gully
集水口 gulley
排水沟 drainage trench
沟盖板 trench cover
活动盖板 removable cover plate
集水坑 sump pit
孔翻边 hole up stand
电缆沟 cable trench

10. Doors, Glass, Windows and Hardware 门、玻璃、窗及五金件

木(钢)门 wooden (steel) door
镶板门 paneled door
夹板门 plywood door
铝合金门 aluminum alloy door
卷帘门 roller shutter door

弹簧门 swing door
推拉门 sliding door
平开门 side-hung door
折叠门 folding door
旋转门 revolving door
玻璃门 glazed door
密闭门 air-tight door
保温门 thermal insulating door
镀锌铁丝网门 galvanized steel wire mesh door
防火门 fire door
（大门上的）小门 wicket
门框 door frame
门扇 door leaf
门洞 door opening
结构开洞 structural opening
单扇门 single door
双扇门 double door
疏散门 emergency door
纱门 screen door
门槛 door sill
门过梁 door lintel
上冒头 top rail
下冒头 bottom rail
门边木 stile
槽口 notch
木窗 wooden window
钢窗 steel window
铝合金窗 aluminum alloy window
百叶窗（通风为主）sun-bind, louver (louver, shutter, blind)
塑钢窗 plastic steel window

空腹钢窗 hollow steel window
固定窗 fixed window
平开窗 side-hung window
推拉窗 sliding window
气窗 transom
上悬窗 top-hung window
中悬窗 center-pivoted window
下悬窗 hopper window
活动百叶窗 adjustable louver
天窗 skylight
老虎窗 dormer window
密封双层玻璃 sealed double glazing
钢筋混凝土过梁 reinforced concrete lintel
钢筋砖过梁 reinforced brick lintel
窗扇 casement sash
窗台 window sill
窗台板 window board
窗帘盒 curtain box
合页(铰链) hinge (butts)
转轴 pivot
长脚铰链 parliament hinge
闭门器 door closer
地弹簧 floor closer
插销 bolt
门锁 door lock
拉手 pull
链条 chain
门钩 door hanger
碰球 ball latch
窗钩 window catch

暗插销 insert bolt
电动开关器 electric opener
平板玻璃 plate glass
夹丝玻璃 wire glass
透明玻璃 clear glass
毛玻璃（磨砂玻璃） ground glass (frosted glass)
防弹玻璃 bullet-proof glass
石英玻璃 quartz glass
吸热玻璃 heat absorbing glass
磨光玻璃 polished glass
着色玻璃 pigmented glass
玻璃瓦 glass tile
玻璃砖 glass block
有机玻璃 organic glass

11. Staircase, Landing & Lift (Elevator) 楼梯、休息平台及电梯

楼梯 stair
楼梯间 staircase
疏散梯 emergency stair
旋转梯 spiral stair (circular stair)
吊车梯 crane ladder
直爬梯 vertical ladder
板式楼梯 cranked slab stairs
踏步 step
扇形踏步 winder (wheel step)
踏步板 tread
挡步板 riser
踏步宽度 tread width
防滑条 non-slip insert (strips)
栏杆 railing (balustrade)

平台栏杆 platform railing
吊装孔栏杆 railing around mounting hole
扶手 handrail
梯段高度 height of flight
防护梯笼 protecting cage (safety cage)
平台 landing (platform)
操作平台 operating platform
装卸平台 platform for loading & unloading
楼梯平台 stair landing
客梯 passenger lift
货梯 goods lift
客/货两用梯 goods/passenger lift
液压电梯 hydraulic lift
自动扶梯 escalator
观光电梯 observation elevator
电梯机房 lift mortar room
电梯坑 lift pit
电梯井道 lift shaft

12. Roofing and Celling 屋面及天棚

女儿墙 parapet
雨篷 canopy
屋脊 roof ridge
坡度 slope
坡跨比 pitch
分水线 water-shed
附加油毡一层 extra ply of felt
檐口 eave
挑檐 overhanging eave
檐沟 eave gutter

平屋面 flat roof
坡屋面 pitched roof
雨水管 downspout, rain water pipe (R. W. P)
泛水 flashing
内排水 interior drainage
外排水 exterior drainage
滴水 drip
屋面排水 roof drainage
找平层 leveling course
卷材屋面 built-up roofing
天棚 ceiling
檩条 purlin
屋面板 roofing board
天花板 ceiling board
防水层 water-proof course
检查孔 inspection hole
上人孔 manhole
吊顶 suspended ceiling, false ceiling

13. Building Material 建筑材料

(1) Bricks and Tiles 砖和瓦
  红砖 red brick
  黏土砖 clay brick
  瓷砖 glazed brick (ceramic tile)
  防火砖 fire brick
  空心砖 hollow brick
  面砖 facing brick
  地板砖 flooring tile
  马赛克 mosaic
  陶粒混凝土 ceramsite concrete

琉璃瓦 glazed tile
脊瓦 ridge tile
石棉瓦 asbestos tile (shingle)
波形石棉水泥瓦 corrugated asbestos cement sheet
(2) Lime, Sand and Stone 灰、砂和石
石膏 gypsum
大理石 marble
汉白玉 white marble
花岗岩 granite
碎石 crushed stone
毛石 rubble
蛭石 vermiculite
珍珠岩 perlite
水磨石 terrazzo
卵石 cobble
砾石 gravel
粗砂 course sand
中砂 medium sand
细砂 fine sand
(3) Cement and Mortar 水泥、砂浆
波特兰水泥(普通硅酸盐水泥) Portland cement
硅酸盐水泥 silicate cement
普通硅酸盐水泥 ordinary Portland cement
火山灰水泥 pozzolan cement
白水泥 white cement
水泥砂浆 cement mortar
石灰砂浆 lime mortar
水泥石灰砂浆(混合砂浆) cement-lime mortar
保温砂浆 thermal mortar
防水砂浆 water-proof mortar

耐酸砂浆 acid-resistant mortar
耐碱砂浆 alkaline-resistant mortar
沥青砂浆 bituminous mortar
纸筋灰 paper strip mixed lime mortar
麻刀灰 hemp cut lime mortar
灰缝 mortar joint

(4) Facing and Plastering Materials 饰面及粉刷材料
水刷石 granitic plaster
斩假石 artificial stone
刷浆 lime wash
可赛银 casein
大白浆 white wash
麻刀灰打底 hemp cuts and lime as base
喷大白浆两道 sprayed twice with white wash
分格抹水泥砂浆 cement mortar plaster sectioned
板条抹灰 lath and plaster

(5) Asphalt (Bitumen) and Asbestos 沥青和石棉
沥青卷材 asphalt felt
沥青填料 asphalt filler
沥青胶泥 asphalt grout
冷底子油 adhesive bitumen primer
沥青玛啼脂 asphaltic mastic
石棉板 asbestos sheet
石棉纤维 asbestos fiber

(6) Timber 木材
裂缝 crack
透裂 split
环裂 shake
干缩 shrinkage
翘曲 warping

原木 log
圆木 round timber
方木 square timber
板材 plank
木条 batten
板条 lath
木板 board
红松 red pine
白松 white pine
落叶松 deciduous pine
云杉 spruce
柏木 cypress
白杨 white poplar
桦木 birch
冷杉 fir
栎木 oak
榴木 willow
榆木 elm
杉木 cedar
柚木 teak
樟木 camphor wood
防腐处理的木材 preservative-treated lumber
胶合板 plywood
三(五)合板 3(5)-plywood
企口板 tongued and grooved board
层夹板 laminated plank
胶合层夹木材 glue-laminated lumber
纤维板 fiber-board
竹子 bamboo

(7) Metallic Materials 金属材料

圆钢 steel bar
方钢 square steel
扁钢 steel strap, flat steel
型钢 steel section (shape)
槽钢 channel
角钢 angle steel
等边角钢 equal-leg angle
不等边角钢 unequal-leg angle
工字钢 I-beam
宽翼缘工字钢 wide flange I-beam
T 型钢/丁字钢 T-stell
冷弯薄壁型钢 light gauge cold-formed steel shape
热轧 hot-rolled
冷轧 cold-rolled
冷拉 cold-drawn
冷压 cold-pressed
合金钢 alloy steel
钛合金 titanium alloy
不锈钢 stainless steel
竹节钢筋 corrugated steel bar
变形钢筋 deformed bar
光圆钢筋 plain round bar
钢板 steel plate
薄钢板 thin steel plate
低碳钢 low carbon steel
冷弯 cold bending
钢管 steel pipe (tube)
无缝钢管 seamless steel pipe
焊接钢管 welded steel pipe
黑铁管 iron pipe

镀锌钢管 galvanized steel pipe
铸铁 cast iron
生铁 pig iron
熟铁 wrought iron
镀锌铁皮 galvanized steel sheet
镀锌铁丝 galvanized steel wire
钢丝网 steel wire mesh
多孔金属网 expanded metal
锰钢 manganese steel
高强度合金钢 high strength alloy steel
有色金属 non-ferrous metal
黑色金属 ferrous metal
金 gold
白金 platinum
铜 copper
黄铜 brass
青铜 bronze
银 silver
铝 aluminum
铅 lead

(8) Anti-corrosion Materials 防腐蚀材料
聚乙烯 polythene, polyethylene
尼龙 nylon
聚氯乙烯 PVC (polyvinyl chloride)
聚碳酸酯 polycarbonate
聚苯乙烯 polystyrene
丙烯酸树脂 acrylic resin
乙烯基酯 vinyl ester
橡胶内衬 rubber lining
氯丁橡胶 neoprene

沥青漆 bitumen paint
环氧树脂漆 epoxy resin paint
氧化锌底漆 zinc oxide primer
防锈漆 anti-rust paint
耐酸漆 acid-resistant paint
耐碱漆 alkali-resistant paint
水玻璃 sodium silicate
树脂砂浆 resin-bonded mortar
环氧树脂 epoxy resin

(9) Building Hardware 建筑五金
钉子 nails
螺纹屋面钉 spiral-threaded roofing nail
环纹石膏板钉 annular-ring gypsum board nail
螺丝 screws
平头螺丝 flat-head screw
螺栓 bolt
普通螺栓 commercial bolt
高强螺栓 high strength bolt
预埋螺栓 insert bolt
胀锚螺栓 cinch bolt
垫片 washer

(10) Paint 油漆
底漆 primer
防锈底漆 rust-inhibitive primer
防腐漆 anti-corrosion paint
调和漆 mixed paint
无光漆 flat paint
透明漆 varnish
银粉漆 aluminum paint
磁漆 enamel paint

干性油 drying oil
稀释剂 thinner
焦油 tar
沥青漆 asphalt paint
桐油 Tung oil, Chinese wood oil
红丹 red lead
铅油 lead oil
腻子 putty

14. Structure Engineering 结构工程

混合结构 mixed structure
板柱结构 slab-column system
拱结构 arch structure
折板结构 folded-plate structure
壳体结构 shell structure
风架结构 space truss structure
悬索结构 cable-suspended structure
框架—剪力墙结构 frame-shear wall structure
筒体结构 tube structure
高耸结构 high-rise structure
框架结构 frame structure
剪力墙结构 shear-wall systems
钢筋混凝土结构 reinforced concrete structure
现浇钢筋混凝土结构 cast-in-place reinforced concrete
无梁楼盖 flat slab
徐变 creep
梁 beam
柱 column
板 slab
剪力墙 shear wall

剪力 shear
剪切变形 shear deformation
剪切模量 shear modulus
拉力 tension
压力 pressure
延伸率 percentage of elongation
位移 displacement
应力 stress
应变 strain
应力集中 concentration of stresses
应力松弛 stress relaxation
应力图 stress diagram
应力应变曲线 stress-strain curve
应力状态 state of stress
加载 loading
抗压强度 compressive strength
抗弯强度 bending strength
抗拉强度 tensile strength
裂缝 crack
屈服 yield
屈服点 yield point
屈服荷载 yield load
屈服极限 limit of yielding
屈服强度 yield strength
屈服强度下限 lower limit of yield
荷载 load
横截面 cross section
承载力 bearing capacity
承重结构 bearing structure
弹性模量 elastic modulus

预应力钢筋混凝土 prestressed reinforced concrete
预应力钢筋 prestressed reinforcement
预应力损失 loss of prestress
预制板 precast slab
主梁 main beam
次梁 secondary beam
弯矩 moment
悬臂梁 cantilever beam
延性 ductility
塑性 plasticity
刚度 rigidity
脆性 brittleness
脆性破坏 brittle failure
受弯构件 member in bending
受拉区 tensile region
受压区 compressive region
轴向压力 axial pressure
轴向拉力 axial tension
吊车梁 crane beam
可靠性 reliability
黏结力 cohesive force
外力 external force
偏心受拉 eccentric tension
偏心受压 eccentric compression
偏心距 eccentric distance
疲劳强度 fatigue strength
偏心荷载 eccentric load
跨度 span
跨高比 span-to-depth ratio
跨中荷载 midspan load

集中荷载 concentrated load
分布荷载 distribution load
挠度 deflection
设计荷载 design load
设计强度 design strength
构造 construction
简支梁 simple beam
截面面积 area of section
刚架 rigid frame
弯曲破坏 bending failure

15. Civil Engineering Machinery 土木工程机械

塔吊 tower crane
打桩机 pile drive
铲动机 scraper
抓斗式挖土机 clamshell
拉索挖土机 dragline excavator
挖土机 backhoe
推土机 bulldozer
铲土机 shovel loader
平路机,压路机 motor grader
挖泥船 dredger
轮胎式压路机 tire roller
三轮压路机 macadam roller
两轮压路机 tandem roller
振动式压路机,夯 vibrating roller
压实机,夯具 compactor/soil-compactor
蛤蟆夯 rammer
沥青喷洒 asphalt sprayer
制造沥青装置 asphalt plant

制造混凝土装置 concrete plant
混凝土搅拌机 concrete mixer
混凝土搅拌车 truck mixer/concrete mixing car
混凝土泵 concrete pump
振捣器,振动器 vibrator

16. Feasibility Study 可行性研究

可行性研究报告 FSR(feasibility study report)
现场选定 site selection
(现场)位置 location
可获利润 profitability
生产成本 production cost
经营费 operation cost
可变成本 variable cost
不变成本 fixed cost
总投资 total investment cost
流动资本 working capital
固定资本 fixed capital cost
总资本 total capital cost
折旧费 depreciation
现金流量 cash flow
折现现金流量 discount cash flow
工况研究 case study
建设资本利息 interest during construction

17. Tendering and Bidding 招投标

投标邀请书 Invitation to Bid
投标押金,押标金 Bid Bond
投标文件 tender documents
投标书 Form of Tender

投标评估 evaluation of bids
资格预审 prequalification
询价 inquiry
询价请求,询价单 Requisition for Inquiry
报价 quotation
报价表格 Form of Quotation
提交报价 submission of quotation
提价 escalation
标底 base price limit on bids
报标 bid quotation
评标 evaluation of tender
议标 tender discussion
决标 tender decision
开标 bid opening
中标者 the winning/successful bidder/ the successful tenderer
未中标者 the unsuccessful tenderer
中标函 Letter of Acceptance

Appendix 2

建筑工地常用标识语

对施工期间带来的不便表示歉意。Apologize for any inconvenience caused during building operation.

生命危险,严禁入内。Danger of death, Keep out.

工地危险,禁止入内。Danger, building site, keep out.

危险,请走开。Danger, evacuation.

危险结构,该桥不安全。Dangerous structure, this bridge is unsafe.

正在施工。Hot work in progress.

任何人不许越过此处。No persons allowed beyond this point.

穿安全靴。Safety footwear.

此工地必须戴安全帽。Safety helmets must be worn on this site.

工地入口,危险。Site entrance, dangerous.

工地入口请慢行。Slow, site entrance.

该按钮已卸下拿去修理。This button has been moved for remedial work.

仅供施工人员使用。This is just for construction personnel.

此电梯仅供施工人员使用。This lift is only for construction personal.

上面在施工。Working overhead.

Reference

[1] 孙爱荣,刘晚成.建筑工程专业英语[M].哈尔滨:哈尔滨工业大学出版社,1997.
[2] 新英汉建筑工程词典[M]. 2版.北京:中国建筑工业出版社,2008.
[3] 土木建筑工程英汉词典[M].北京:中国水利水电出版社,2008.